RUSSIAN-DESIGNED REACTORS

HEARING

BEFORE THE

SUBCOMMITTEE ON
ENERGY AND POWER

OF THE

COMMITTEE ON
ENERGY AND COMMERCE
HOUSE OF REPRESENTATIVES

ONE HUNDRED THIRD CONGRESS

FIRST SESSION

OCTOBER 28, 1993

Serial No. 103–86

Printed for the use of the Committee on Energy and Commerce

U.S. GOVERNMENT PRINTING OFFICE
78-866 WASHINGTON : 1994

For sale by the U.S. Government Printing Office
Superintendent of Documents, Congressional Sales Office, Washington, DC 20402
ISBN 0-16-044259-1

COMMITTEE ON ENERGY AND COMMERCE

JOHN D. DINGELL, Michigan, *Chairman*

HENRY A. WAXMAN, California
PHILIP R. SHARP, Indiana
EDWARD J. MARKEY, Massachusetts
AL SWIFT, Washington
CARDISS COLLINS, Illinois
MIKE SYNAR, Oklahoma
W.J. "BILLY" TAUZIN, Louisiana
RON WYDEN, Oregon
RALPH M. HALL, Texas
BILL RICHARDSON, New Mexico
JIM SLATTERY, Kansas
JOHN BRYANT, Texas
RICK BOUCHER, Virginia
JIM COOPER, Tennessee
J. ROY ROWLAND, Georgia
THOMAS J. MANTON, New York
EDOLPHUS TOWNS, New York
GERRY E. STUDDS, Massachusetts
RICHARD H. LEHMAN, California
FRANK PALLONE, JR., New Jersey
CRAIG A. WASHINGTON, Texas
LYNN SCHENK, California
SHERROD BROWN, Ohio
MIKE KREIDLER, Washington
MARJORIE MARGOLIES-MEZVINSKY, Pennsylvania
BLANCHE M. LAMBERT, Arkansas

CARLOS J. MOORHEAD, California
THOMAS J. BLILEY, JR., Virginia
JACK FIELDS, Texas
MICHAEL G. OXLEY, Ohio
MICHAEL BILIRAKIS, Florida
DAN SCHAEFER, Colorado
JOE BARTON, Texas
ALEX MCMILLAN, North Carolina
J. DENNIS HASTERT, Illinois
FRED UPTON, Michigan
CLIFF STEARNS, Florida
BILL PAXON, New York
PAUL E. GILLMOR, Ohio
SCOTT KLUG, Wisconsin
GARY A. FRANKS, Connecticut
JAMES C. GREENWOOD, Pennsylvania
MICHAEL D. CRAPO, Idaho

ALAN J. ROTH, *Staff Director and Chief Counsel*
DENNIS B. FITZGIBBONS, *Deputy Staff Director*
MARGARET A. DURBIN, *Minority Chief Counsel and Staff Director*

SUBCOMMITTEE ON ENERGY AND POWER

PHILIP R. SHARP, Indiana, *Chairman*

EDWARD J. MARKEY, Massachusetts
RICHARD H. LEHMAN, California
CRAIG A. WASHINGTON, Texas
MIKE KREIDLER, Washington
BLANCHE M. LAMBERT, Arkansas
AL SWIFT, Washington
MIKE SYNAR, Oklahoma
W.J. "BILLY" TAUZIN, Louisiana
RALPH M. HALL, Texas
RICK BOUCHER, Virginia
JIM COOPER, Tennessee
JOHN D. DINGELL, Michigan (Ex Officio)

MICHAEL BILIRAKIS, Florida
JOE BARTON, Texas
J. DENNIS HASTERT, Illinois
CLIFF STEARNS, Florida
SCOTT KLUG, Wisconsin
GARY A. FRANKS, Connecticut
MICHAEL D. CRAPO, Idaho
CARLOS J. MOORHEAD, California (Ex Officio)

SHELLEY N. FIDLER, *Acting Staff Director*
SUE D. SHERIDAN, *Counsel*
CATHERINE G. VAN WAY, *Minority Counsel*

CONTENTS

RUSSIAN-DESIGNED REACTORS

THURSDAY, OCTOBER 28, 1993

HOUSE OF REPRESENTATIVES,
COMMITTEE ON ENERGY AND COMMERCE,
SUBCOMMITTEE ON ENERGY AND POWER,
Washington, DC.

The subcommittee met, pursuant to notice, at 10:07 a.m., in room 2322, Rayburn House Office Building, Hon. Philip R. Sharp (chairman) presiding.

Mr. SHARP. The subcommittee will please come to order.

I am pleased to welcome three distinguished administration witnesses to our hearing today to discuss the United States' effort to upgrade the safety of Soviet-designed nuclear power reactors.

Nearly 60 of these reactors are operating in the former Soviet Union and Eastern European nations, including two at the Chernobyl plant, the site of, of course, the tragic accident in 1986.

Some of these designs sprung from the Soviet military program, none of the plants were designed to meet Western safety standards, and one of the important questions that I hope to gain a better understanding of today is the degree to which the various types of reactors can be upgraded to operate within acceptable risk limits.

From the standpoint of public risk, it seems clear that at least some of these reactors should be retired immediately. Nonetheless in terms of economics and energy supply, it is equally clear that, at present, the host states' reliance on these facilities seems to be substantial.

One of the questions I am interested in is how the United States can help these countries develop energy alternatives so that they will be in a position to retire the riskiest of the reactors.

Finally, it is critical that the United States' assistance be effective, that it address the most serious problems at Soviet-designed plants first, and that we be assured that our contribution is meaningful.

This goal poses several prerequisites: that we understand the risks, that we provide aid that is meaningful, and that we maintain confidence within our own government that the dollars are well accounted for.

I know Secretary O'Leary is strongly committed to improving departmental and contractor accountability and I urge her to extend that effort to the Soviet reactor initiative as well.

Of course, the overwhelming responsibility rests with the host states with whom we can only hope to work. We cannot force them to take actions that might be of critical importance both to their people and to their neighbors in Europe.

With that, let me recognize the distinguished gentleman from Massachusetts.

Mr. MARKEY. I thank you, Mr. Chairman, and I applaud your holding of this hearing.

I join you in welcoming Secretary O'Leary; and the Chairman of the Nuclear Regulatory Commission, Mr. Selin; and Ms. Verville to the subcommittee this morning, and I would like to commend you again for holding the hearing on this important topic.

I remember quite vividly a hearing which was conducted in this room just days after the news of the Chernobyl nuclear accident first broke to hear dramatic testimony from the Soviet Embassy Representative Vitaly Cherkin—sitting right where you are, Madam Secretary—regarding the health and environmental consequences of the disaster which was unfolding in the Soviet Union.

It is now more than 6 years later and the Soviet Union no longer exists. Gorbachev is no longer in power. Communism is dead in Russian and the former Soviet republics. But the consequences of that fateful event still persist. Hundreds of square miles of territory surrounding the Chernobyl plant remain contaminated and continue to threaten the long-term health of the people who live in that region of the world.

Since the time of the Chernobyl disaster, the U.S. Department of Energy, the Nuclear Regulatory Commission and the State Department have worked to provide health and safety assistance to Russia and the former Soviet republics.

I look forward to hearing the testimony this morning on the status of those efforts and your future plans. At the same time, I believe that as we consider the issue of nuclear safety in Russia and the republics, we cannot limit ourselves solely to the question of whether the operational and safety standards at the reactors are sufficient to avoid another Chernobyl. We must also consider what is being done to assure that the Russians cease all production of fissile materials for nuclear weapons. At the same time, we must have some assurance that weapons, usable nuclear materials produced as a result of civilian reactor operations are properly safeguarded, and that such materials are not used or diverted for nuclear weapons purposes.

I will be very interested in hearing from our witnesses what steps their respective agencies are undertaking to respond to the proliferation challenge.

In addition, while I strongly support immediate efforts to respond to the pressing safety problems affecting the aging and deteriorating reactors, it would be short-sighted for America to adopt a "no dough unless it glows" energy assistance policy for Russia and the republics. I will, therefore, be very eager to hear from our witnesses exactly what they are doing to promote conservation, energy efficiency and alternative energy strategies in Russia and the republics so that they may reduce their dependence on dangerous nuclear reactor technologies.

I thank the witnesses for joining us this morning and I yield back, Mr. Chairman, the balance of my time.

Mr. SHARP. We are very pleased now to recognize our witnesses.

First, we have the Honorable Hazel R. O'Leary, the Secretary of the United States Department of Energy. We also have Ms. Elizabeth Verville, the Senior Coordinator for Nuclear Reactor Safety and the International Science Center with the United States Department of State, and we are very pleased to have the very Honorable Ivan Selin, the Chairman of the Nuclear Regulatory Commission.

You are all familiar with our process. We will be happy to recognize you for your statement and will be happy to make part of our printed record any printed materials you have.

So Secretary O'Leary, we will be happy to begin with you.

STATEMENTS OF HON. HAZEL R. O'LEARY, SECRETARY, DEPARTMENT OF ENERGY; ELIZABETH VERVILLE, SENIOR COORDINATOR FOR REACTOR SAFETY AND INTERNATIONAL SCIENCE AND TECHNOLOGY CENTERS, DEPARTMENT OF STATE; AND HON. IVAN SELIN, CHAIRMAN, NUCLEAR REGULATORY COMMISSION

Secretary O'LEARY. Thank you, Mr. Chairman.

My formal statement has been provided to this committee as well as attachments which respond in detail to the issues you raised. First of all, with respect to budget and other questions that have been either jointly or singularly answered by this group, I would just have a few comments.

Mr. Markey, I, too, have some vivid recollections going back to a time when you and I were a lot younger. My first vivid recollection involves conversations with my colleagues at the Department of Energy back in 1979, when a group of some Cabinet appointees went over to Russia and toured many of their reactors. A vivid recollection of this was that there are accidents waiting to happen there. Of course, when Chernobyl shocked the world, all of us understood that the concerns of those who in the Department of Energy, went on what must have been the first tour of some of those reactors for many.

You know and I know the history that is woven from the Chernobyl incident, and I won't recite it, since it is a part of my formal testimony. I do want to touch upon a few issues, and I will do them very quickly, too.

We have had our people over in Russia since 1986, '87 and '88, working first in a very collaborative way, and finally in a much more structured and formal way on issues involving operation management, systems design efficiencies, fire and safety. The DOE framework, I believe, has been much strengthened since the Commission meetings with Gore and Chernomyrdin, which grew out of Vancouver and, subsequently, out of the Tokyo Summit.

Part of the complex framework that supports our effort to work on the issues of safety in the former Soviet states regarding these nuclear power plants involves itself, of course, not only in the new leadership coming out of the Gore/Chernomyrdin Commission, where there is a specific subgroup working on issues involving nuclear safety co-chaired by Chairman Selin and myself, but, too, we thread the international cooperation that comes out of the International Atomic Energy Agency. I will subsequently discuss, if asked, visits to Vienna just last month, as well as Moscow, where we focused on some of the issues that have been raised by both the chairman and Mr. Markey.

I, too, recognize that with the work that has been ongoing with INPO, as well as the Department of Energy and our laboratories, what seems to be missing from that work is pronouncement of and laying out of metrics, which will indicate to the public, as well as this committee and others who are interested, what progress is being made on work now ongoing.

I have had discussions with Hans Blix at the International Atomic Energy Agency and indicated to him that I believe the set of metrics were important, not just in the United States but internationally as well. My formal testimony points out that he, too, agrees, and that the International Atomic Energy Agency has determined that they need to establish a set of metrics that can follow progress.

Those performance measures, with respect to the work that the Department of Energy has undertaken, I can assure you will be developed and shared with this committee in a very formal way as we move forward with the fiscal '95 budget and hopeful, as soon as this hearing is concluded. That is not to indicate that there have not been performance measures, but I don't believe that they have been carefully articulated nor are we, as a public, tracking the progress we are making against moneys expended.

With respect to the overall assistance program, I want to point out that this committee and the Congress generally were involved in the recent provision of approximately $100 million to focus on the safety initiatives that have now been ongoing and sharpened most recently to involve the work taking place in Russia as well as the former Soviet Union and in Central and Eastern Europe.

I will be happy to respond to questions specifically with respect to the breakdown of those funds, but tell you that we are still very much in negotiation, both with state and with some of our contracts with respect to how those moneys should be spent.

That permits me now to raise a very sticky problem having to do with liability coverage, which has been a major concern to both private companies, who have been working in partnership with Russian and other former Soviet State utilities, as well as working as contractors. Most of the work that has been ongoing in the area of both operation and fire safety has been on-track and relatively on-budget. The difficulty that now faces us, as we further attempt to negotiate acceptance of liability in the former Soviet States, is the strong signal from our own contractors and private sector partners that unless liability issues are resolved to their satisfaction, work that is now planned through January and into the next year will not be undertaken. I take this as a very serious issue.

My general counsel, Bob Nordhaus, is meeting with some of these contractors even today as we sit here, and I am sure we will have more discussion on that as we submit to question and answers. I consider this a critical issue. I am pleased with the fact that some work is going forward and suggest to you that we will likely be talking to one another as we go forward with it.

I want to briefly touch on the challenge of U.S. leadership and point out that some I know have felt that there has been a lack of high level attention to these issues in the prior administration. The question has been raised whether this high level review is taking place in the Clinton Administration.

I would like to leave this committee with one thought, that, as a result of the leadership of the President and the Vice President exercised through the Gore/Chernomyrdin Commission, my sense is that coordination is at a very high level. I can't speak for what happened in the prior administration, but I can tell you that the work of Gore/Chernomyrdin Commission is reviewed and discussed on a very frequent basis with the Vice President of the United States. As a matter of fact it was discussed with the Vice President last Friday. So, if there remains some lingering concern that there is no high level attention to this issue, I want to make it clear that there is attention at the very highest level.

I would touch for just a moment on the issues raised by Mr. Markey with respect to energy efficiency and what work is ongoing on the part of this administration in the former Soviet Union. Again, out of the Gore/Chernomyrdin Commission Meetings and subsequent to that, there have been activities both in Russia and the Ukraine and in Eastern Europe focusing on energy efficiency, grants let, and I would specifically speak to the Energy Efficiency Center that is now in operation in Moscow.

I would further point out to you, which I think is in my testimony, the fact that the Prime Minister, Chernomyrdin himself, at the last commission meeting stated quite without prompting from any of us that in his belief, energy efficiency was an important driver for the Russian economy. He expressed the idea that he, having worked in Gasprom, understood the fact that a kilowatt saved is much cheaper than one generated. The recent memorandum of cooperation signed with the Russian Federation will permit us to establish additional energy efficiency centers in the Russian Federation and to build on the work that was done there.

In addition, the plan for the beginning of Fiscal Year '95, if not sooner, is to begin a relighting effort, first of all in Moscow proper, which should save approximately 50 percent of the energy used in lighting the entire city of Moscow. That is to be done with funds created in a revolving fund not unlike the prototypes that have been established in the United States.

I am also pleased to indicate to you that U.S. Wind Power has entered into a joint agreement in the Ukraine for alternative energy. So there are these examples of projects now up and working, and the agreements that were signed between the United States and the Russian Federation in September that provide the groundwork for other projects which will permit us to build on the earlier successes.

It is clear to me, as it was to my Russian counterparts when I met with them last month in Moscow, that the opportunity for energy efficiency is one that should be seized upon. This is understood by the Minister of Energy, as well as the Minister of Ecology, by the President's Chief Environmental Advisor, Alexi Yablakov. Perhaps most importantly, because I believe that this is new ground uncovered, when I was in Moscow last month, I met with a number of the leaders of Russian non-governmental organizations focusing on issues involving energy efficiency, ecology, and those who are also concerned about issues of clean-up at the nuclear production sites.

I will tell you that there is there a groundswell of public interest to be energized on this issue, and one of the things that we committed to do when I return to Russia, perhaps in December, is to ensure that there is an international conference via satellite hookup, if at all possible, between organizations in the United States who are interested in these issues and their Russian counterparts.

I believe that with both the funding and the focus of the Gore/Chernomyrdin Commission, and a new and energized group of public citizens interested in these issues, we can provide the groundwork to move forward on energy efficiency issues.

My sense is that unless we are working there as well as in areas of natural resource exploration, we are not going to provide the economic wherewithal to permit us, and to permit the Russians, to move off of this dependence on nuclear power that is generated both with the older type reactors and at their nuclear weapons production sites. I believe that the pressure is there to get this work done.

I commend this committee for this hearing, and point out to you that I think a strong dose of a hearing like this for us in the administration is also a very good idea and I welcome these opportunities and look forward to answering your questions.

[Testimony resumes on p. 26.]

[The prepared statement of Ms. O'Leary follows:]

STATEMENT OF
HAZEL R. O'LEARY
SECRETARY OF ENERGY

Mr. Chairman and members of the Committee, thank you for this opportunity to discuss the Department of Energy's nuclear safety assistance programs focused on Soviet-designed reactors in Russia, Ukraine and the Central and Eastern European Countries.

As a former executive of a utility operating both fossil and nuclear powerplants, I gained experience with an integrated electrical generation system and the culture needed for safe operation of nuclear plants. This experience will be helpful in our efforts to assist countries of the former Soviet Union find alternatives to the operation of the older, less safe reactors and to instill an effective safety culture for reactor operation.

In this testimony I will review our current programs and examine some of the challenges we face in our nuclear safety assistance. Additionally, I will explain the important coordination role that the Gore/Chernomyrdin Commission -- established by President Clinton and President Yeltsin during the Vancouver summit -- is playing on this issue.

While at DOE in 1979, I vividly remember a Department team visiting a Soviet RBMK reactor and reporting that an accident was "waiting to happen." Unfortunately, they were proved correct with the accident in 1986 at the Chernobyl plant. However, as is often the case, tragic events prompt needed actions.

In 1986, in an effort to understand the Chernobyl accident better, a team of DOE experts analyzed its causes and consequences. Their November 1986 report confirmed that the power excursion was caused by serious violations of procedures by plant operators and fundamental flaws in plant design.

The former Soviet Union also designed and constructed another type of nuclear power plant, referred to as the VVER, which is similar to Pressurized Water Reactors in use in Western countries. Two DOE reports in 1987 and 1989 analyzed the safety-related features of VVERs as well as the probable behavior of these units under a variety of accident conditions. The reports were provided to the Russians and many parties in the West and have been helpful in our current initiatives.

As Chernobyl showed, a major nuclear accident anywhere touches us all. Just as we know that the safety and environmental impact of our own energy production is paramount, we also need to understand that energy production in other nations must have the same emphasis. This awareness is especially critical as regards the countries of the former Soviet Union. Their energy systems stand to gain great benefits in the exchange of information with, and experience from, U.S. industry.

With funding from the Agency for International Development, the Energy Department has a growing program of energy assistance to former Soviet Union countries. This assistance program is designed to:

- o encourage energy efficiency and energy supply and demand improvements that, among other things, strengthen the economy and allow the shutdown of the RBMK and older VVER reactors,
- o assist with environmental restoration, and
- o support reactor safety through hardware upgrades, improved safety culture, and a strong and independent regulator.

Our assistance must be tied to the economic security of these countries. We seek fundamental reform of the energy sectors to enhance energy security and consequently to improve the nuclear safety situation. It is vital that the U.S. help countries of the former Soviet Union face one of their greatest challenges, which is to find the economic and technical wherewithal to support alternatives to operation of the more dangerous reactors. For example, in Russia we are studying ways to decrease dependence on the nation's older reactors such as increased gas use, life extension of their fossil fuel stations, power wheeling and more efficient load dispatching.

Critical to U.S. and European efforts to reduce reactor risks and promote alternative energy sources will be the ability to measure our performance. The U.S. and western European nations have made substantial financial commitments to this serious problem and we must be able to measure our progress accurately. In a recent meeting, the Director General of the International Atomic Energy Agency (IAEA), Hans Blix, echoed the critical importance of performance measures in U.S. and European nuclear safety assistance.

I. DOE NUCLEAR SAFETY INITIATIVES

In the post-Chernobyl era, the U.S. instituted a wide-ranging program of cooperation with the then Soviet Union to improve reactor safety, particularly with respect to the RBMKs and older model VVERs. I will briefly describe each element of the program and touch on related performance measures.

A. Operational Safety Initiative at Russia's Novovoronezh Reactors

Our initial emphasis, begun in 1990, was to help improve operational safety and response to emergency conditions for the VVER-440/Model-230 reactors. This effort was led by the Institute for Nuclear Power Operations and U.S. utilities and emphasized the lessons learned in the U.S. following the Three Mile Island accident. Emergency procedures and training materials are completed. Training of operators is underway and the use of emergency procedures is to begin at the Novovoronezh plant in southern Russia in early 1994. Other operations management and control improvements such as the addition of shift supervisors have already been implemented at Novovoronezh. The Russian regulatory body, Gosatomnadzor (GAN) is reviewing, and will approve, the implementation of the emergency operating procedures at Novovoronezh.

B. Current Nuclear Safety Assistance Initiative in Russia and Ukraine

At the Group of Seven meeting in May 1992, a five-point plan for safety improvements was adopted which included many elements of our current program:

- o operational safety improvements,
- o nuclear safety regulation support,
- o interim key risk reduction actions for RBMKs and older VVERs,
- o examination of alternative energy requirements to limit the remaining lifetime of RBMK and older VVERs, and
- o needed upgrades for the newer VVER plants.

The current U.S. nuclear safety assistance initiative began in May 1992 and was continued and expanded by President Clinton at the Vancouver summit and again at the Tokyo Summit. The current effort draws on U.S. utility experience, concentrating on specific system upgrades and general safety assistance. It does not, however, include efforts which could be used to extend the lifetimes of RBMKs and older VVERs.

The U.S. initiative has five major elements, the first four of which are the responsibility of DOE:

- o extending operational safety enhancements developed for the VVER-440/Model-230s at Novovoronezh to all other Russian design types,
- o establishing regional training centers in Russia and Ukraine,
- o providing risk reduction measure at selected RBMK and VVER-440/Model-230 plants,
- o providing fire safety upgrades for plants in Russia and Ukraine, and
- o assisting Russian and Ukrainian regulators in training and developing consistent safety standards and procedures (NRC-lead).

A listing of specific projects being undertaken by DOE, with the overall project schedule, is provided as an attachment. AID provided funding of $25 million in fiscal year 1992 and $30 million in fiscal year 1993. The status of each of the elements follows:

1. Operational Safety Enhancements

Operational safety enhancements focus on the development of emergency operating procedures and on the improvement of operational and management controls. Emergency procedures are being developed for initial plants in Russia and Ukraine. Once developed jointly by the Russians and Ukrainians with U.S. assistance, the new emergency operating procedures will be adapted by the Russians and Ukrainians to all of their units. As shown on the attachment, work has begun and actual implementation at the target plants will take place in 1995. The Institute for Nuclear Power Operations (INPO) and seven U.S. utilities are leading this effort for the U.S. We will gauge our success in part through direct and detailed observation of operations and tracking of reported incidents.

2. Training Centers

Two regional training centers are being established by Russia and Ukraine with U.S. assistance. They are to serve as focal points for training and for establishing comprehensive Western safety management systems. The Centers are operated by the Russians at the Balakovo plant and by the Ukrainians at the Khmelnitsky plant. U.S. assistance includes funding for development of 12 key training programs, which are the same as specified by INPO for plants in the U.S. Both of these training centers incorporate the use of plant simulators for operator training. These training centers should become operational in Russia in early 1995 and Ukraine in early 1996. These efforts will be measured through assessments of operator performance and review of reported incidents.

3. Risk Reduction

Another key element of the current nuclear safety assistance program is interim risk reduction actions. These actions are being taken for the highest risk plants to quickly reduce the safety hazards during their remaining lifetime. We will not assist with plant upgrades that could encourage continued plant operation. The following actions reduce the risk from severe accidents while not encouraging long-term operation:

- improve confinement performance for severe accidents through filtered venting and reduced leakage,
- prevent uncontrolled hydrogen explosions in the confinement in the event of a severe accident,
- install a dedicated emergency diesel and feedwater pump,
- assure that basic fire detection and response capabilities are provided at each plant, and
- install automatic reactor trips for remaining high-risk failure modes.

78-866 O - 94 - 2

These risk reduction measures involve U.S. industry working closely with the designers, owners and operators of these plants. Feasibility and conceptual design studies have already been completed and we have contracted with the U.S. private sector to work with their counterparts overseas to prepare detailed plans for implementation of the agreed-to measures. The continuation of this work is tied to the completion of the liability agreements, which I will discuss below. Some of these projects are proceeding at a fast pace, with confinement sealing testing having been successfully performed at the Kola plant in northern Russia last September.

4. Fire Safety

Soviet-designed reactors continue to experience fires. Many of these fires could have been prevented or reduced in severity with adequate fire safety systems. We have provided support for an integrated fire safety program that includes fire detection and mitigation. This initiative is underway with sample hardware already being delivered to the Smolensk nuclear power plant in Russia and the Zaporazhye nuclear power plant in Ukraine. This work is also tied to completion of the liability agreements. The ultimate objective is to transfer our fire protection technology to the Russians and Eastern Europeans to assist them in completing these upgrades at the rest of their Soviet-designed reactors. Success of this effort will be directly measured by a reduction in the number and severity of fires being experienced.

C. Safety Assistance in Central and Eastern European Countries

The DOE program in the Central and Eastern European Countries primarily responds to specific assistance requests from these countries. These requests have involved fire fighting, personnel safety, and emergency equipment. For fiscal year 1994, we are starting to take advantage of the work being done in Russia, especially adaptation of symptom-based emergency operating instructions. However, only limited efforts in areas such as fire safety are possible with current funding.

D. Future Initiatives

In September, Congress passed the Foreign Operations Appropriations Act for fiscal year 1994, which includes a package of assistance with about $100 million planned for nuclear safety for Russia, Ukraine, and the other Newly Independent States. The U.S. Government and the recipient countries are developing a prioritized list of nuclear safety projects to use the new funds. For DOE, this will likely include completion of training and safety system upgrades at the initial plants, extension of fire safety improvements to all plants, and application of safety system upgrades from the initial plants to an additional RBMK and VVER-440/Model-230.

Our initiatives are currently directed only at Soviet-designed civilian nuclear powerplants. We have proposed to the Russians to explore the possible conversion or replacement of their plutonium production reactors at Tomsk and

Krasnoyarsk. These weapons plants are still being operated for local heat and power. We are seeking a response from the Russians to our proposal.

E. Redirecting Nuclear Weapons Scientists

While we are placing a high priority on working with the existing nuclear community in Russia to improve the safety of their older reactors, the collapse of the Soviet Union has raised the possibility of redirecting the talents of highly trained and dedicated weapons scientists and engineers to peaceful activities. Ratification of the November 1992 agreement to establish an International Science and Technology Center in Moscow for this purpose stalled in the Russian Supreme Soviet. Recent events may enhance the opportunity for implementing this critical initiative.

Russia needs to place this as a high national priority in order to continue the transition from weapons production to activities that have peaceful applications and contribute to the continued improvement of the Russian energy infrastructure in an environmentally safe and economically sound manner.

II. CHALLENGES

We face many challenges in our efforts to assist countries of the former Soviet Union improve the safety of their reactors. Our work requires a well coordinated effort both within the U.S. and internationally, and it must be integrated with efforts to deal with overall energy needs in these nations.

A. U.S. Leadership

At the Vancouver summit and the economic summit in Tokyo, the President pledged our economic partnership with Russia as the nation works to achieve economic health. President Clinton and President Yeltsin agreed at Vancouver to establish a joint commission on energy and space cooperation. President Clinton assigned the U.S. chairmanship to Vice President Gore. President Yeltsin appointed Prime Minister Chernomyrdin as the Russian chairman. The Gore/Chernomyrdin Commission, which held its first meeting on September 1 and 2, is the focal point for U.S.-Russian cooperation on energy matters. I chair the Commission's Energy Policy Group and, in that capacity, I recently traveled to Moscow where I signed a Memorandum of Cooperation with Russia on Energy Efficiency that established a framework for bilateral cooperation. Additionally, I am Co-Chair, with Chairman Selin, of the Committee's Nuclear Safety Sub-Group. These groups will have regular meetings to establish and review the goals, milestones, and results of our assistance efforts.

I know members of this Committee have expressed concern regarding the Administration's coordination of the various nuclear safety assistance programs. Let me assure you that Vice President Gore is personally committed to this effort and has made it a high priority of the Gore/Chernomyrdin Commission.

Let me also emphasize that the Department of Energy is coordinating closely with the Department of State which provides the policy and international coordination leadership, the Nuclear Regulatory Commission (NRC) which provides assistance in establishing competent regulating authorities, and the Agency for International Development (AID) which coordinates foreign assistance funding. DOE's role is to provide direct technical and operational safety assistance for the nuclear powerplants thereby improving their safety consciousness, management and equipment. This organization, with the leadership provided by Vice President Gore, is working to implement a successful safety assistance program.

B. International Coordination

The important involvement of many nations and international organizations in various nuclear safety initiatives makes international coordination vital to our program's success. International anxiety about the safety of Soviet-designed nuclear powerplants has led to many bilateral and multilateral initiatives. A Nuclear Safety Committee has been set up in Brussels under the Group of 24 Nations (G-24) to coordinate these different efforts. The Department of Energy and the Nuclear Regulatory Commission, in coordination with the State Department, work in close cooperation with this Committee. The United States also participates in activities conducted by international organizations such as the IAEA, the Nuclear Energy Agency (NEA), the United Nations Economic Commission for Europe, the International Energy Agency (IEA), the World Bank, the European Bank for Reconstruction and Development (EBRD), and the World Association of Nuclear Operators (WANO). Countries receiving assistance are actively involved in ensuring that donor countries do not duplicate safety assistance efforts.

C. Continuing Safety Concerns

While progress has been made to improve the safety of Soviet-designed reactors since the Chernobyl accident, our concern about their operation remains substantial. Due to their lack of containment and emergency core cooling systems, the designs of the Chernobyl-type RBMKs and the older-design VVER-440/Model-230s will always fall far short of Western safety standards. These reactors cannot practically be upgraded to meet contemporary standards for safety and should not be operated any longer than absolutely necessary.

Beyond the design problems, there is still significant risk due to weaknesses in safety culture, such as:

- a lack of a general attitude that safety is the primary objective and responsibility of all,
- a lack of effective operator training and operational procedures,
- a lack of management systems which foster a positive safety culture, and
- production of power given a higher priority than safety.

Additionally, the powerplants suffer from economic and infrastructure problems. The nuclear powerplants have had difficulty in receiving payments for the power they supply. Consequently, managers may be unduly concerned with meeting payroll and continuing operation.

Most Western nations would prefer that the RBMKs and VVER-440/Model-230s shut down in the near term. Unfortunately, because of the perceived electric demand, steam provided to nearby communities and industries, the need to earn hard currency through electricity exports, and the high cost of replacement power, it appears that the countries operating these types of reactors will shut down few, if any, in the near-term.

D. Improving the Total Energy System

U.S. safety assistance is part of a larger effort to help states of the former Soviet Union improve their energy infrastructure in an environmentally safe and economically efficient manner. This will increase the range of energy options, thereby potentially allowing for the shutdown of the less safe reactors over the intermediate-term.

As a result of the Gore/Chernomyrdin Commission, the U.S. has agreed to perform a joint study with Russia on alternative energy options which would lead to production of electricity economically and at significantly lower risk. I consider this work to be among the most critical of our undertakings in Russia over the next few years.

As part of the Gore/Chernomyrdin effort, we have signed agreements of cooperation with Russian officials on fossil energy, and energy efficiency, and we have an agreement pending in the area of radiation safety and health. These agreements, and other agreements with Central and Eastern European Countries, are designed to help these nations tap American technology to improve their total energy systems and provide alternatives to nuclear stations.

For example:

- o As part of a recently signed Memorandum of Cooperation for Fossil Energy we are pursuing projects to develop: an oil and gas technology center; integration of advanced coal-fired power plant systems; technology and information exchange mechanisms; and a privately financed retail gasoline distribution system.
- o We are helping these countries develop a framework of commercial laws and regulations that will encourage private investment in the energy.
- o DOE helped establish the Moscow Energy Efficiency Center in 1990, which has assisted U.S. industry in identifying projects such as lighting the Moscow ring road and relamping Moscow's streets. The contract for these services between a Houston company and Moscow is valued at $80 million.

We have tried, as much as possible, to ensure that the private sector plays an important role in the execution of these agreements.

In related areas, we expect to establish an Environmental Management Center in Moscow to coordinate the transfer of environmental technology. Additionally, under the pending agreement on radiation safety and health, epidemiological and other health studies will be performed at sites such as Chelyabinsk where there have been radiation releases.

E. Liability

A significant obstacle we now face is liability protection for U.S. contractors who perform nuclear safety assistance work. These contractors are concerned that in the event of a nuclear accident, claimants, including those in nearby countries, could seek judgment against them on the grounds they performed some work or service at the plant. Even the cost of a successful legal defense could be very high. These contractors seek application of the same principle that applies in U.S. law and in most other countries, namely, that the operator of the nuclear plant assumes full responsibility for damage or injury resulting from a nuclear accident. We believe this is a valid concern on the part of the U.S. contractors and their full participation cannot be expected without adequate liability protection.

The Government of Ukraine has agreed to hold harmless and indemnify U.S. companies for liability that may be attributed to them for work performed under the safety assistance program. We are still negotiating with Russia for similar protection for U.S. firms. We are hopeful that we can reach agreement on liability issues with Russia in the very near future.

Some of our planned work has already been delayed pending the resolution of this issue and most work, except training centers, would be affected if agreement is not reached by January. The State Department is leading the negotiations on bilateral indemnification agreements to resolve this issue. However, it is important to note that even with the bilateral agreements, some U.S. companies have expressed concern with providing nuclear safety assistance without indemnification from the U.S. Government due to political and financial instability in Russia and Ukraine.

III. CONCLUSION

Mr. Chairman, our attention must be focused on assisting nations of the former Soviet Union and Central and Eastern Europe establish energy policies and a culture of safety in reactor operations, which would lead to decisions to not operate the RBMK and VVER-440/Model-230 plants any longer than is absolutely necessary. At the same time, we must take near-term steps to reduce the risks of their continued operation. An important part of this effort is to improve the safety culture and management practices of the plant operators. Market pricing in these countries is necessary to encourage conservation and to facilitate commercial financing of alternative energy sources. We will know that we are successful if our assistance promotes the near-term safety of Soviet-designed reactors until the economic and energy development of these nations allows them to replace these powerplants. I believe that with the help of Congress and the international community, we can make substantial improvements in reactor safety, and foster sound economic development at the same time.

U.S. DEPARTMENT OF ENERGY

NUCLEAR SAFETY INITIATIVE - RUSSIA

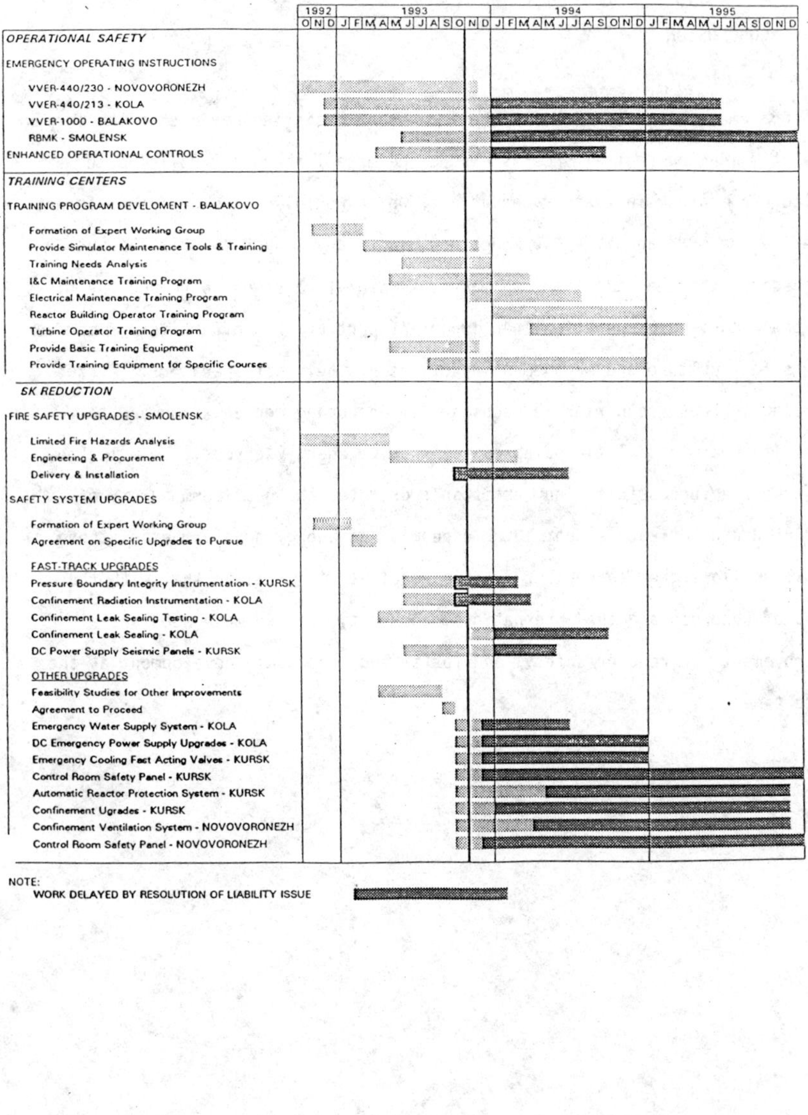

U.S. DEPARTMENT OF ENERGY

NUCLEAR SAFETY INITIATIVE - UKRAINE

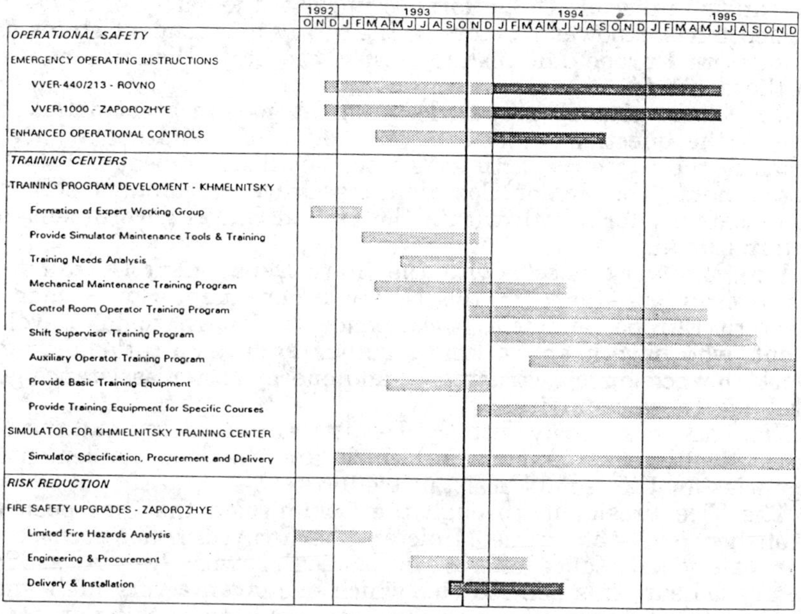

NOTE:
WORK DELAYED BY RESOLUTION OF LIABILITY ISSUE

Mr. SHARP. Thank you very much.

First, Ms. Verville, we are very pleased to hear from you and the State Department at this point.

STATEMENT OF ELIZABETH VERVILLE

Ms. VERVILLE. Thank you, Mr. Chairman, I appreciate this opportunity to discuss U.S. efforts to improve the safety of Soviet-designed nuclear power reactors in the newly independent states and in Eastern Europe, and also to discuss the State Department's role in these efforts.

Mr. Chairman, my prepared statement goes in great detail into one of the questions raised in your letter of invitation, which is, what is the State Department's role. I will not repeat it in great detail here. The Senior Coordinator's Office coordinates U.S. nuclear safety efforts both within the United States government and internationally.

We work very closely with the State Department Coordinators for overall assistance to the Newly Independent States, and to Eastern Europe, and with the Agency for International Development, who have been working together to develop rational frameworks for economic restructuring and energy sector assistance generally in these countries.

We also work closely with the Department of Energy and the Nuclear Regulatory Commission, which are the lead implementing agencies for U.S. safety assistance efforts.

The Vice President, through the Commission that has been established with the Prime Minister Chernomyrdin, has focused on our safety assistance efforts, in particular with Russia, as Secretary O'Leary has pointed out, which has given a very high impetus and boost to the programs that we have been working on.

A great deal has happened in the 7 years that Mr. Markey referred to since the Chernobyl accident. That incident led to U.S. consultations with the former Soviet Union on nuclear safety and the concerns we had about Soviet reactors, including their design, operation and condition. That dialogue grew into limited technical cooperation. But because of overall U.S. relations with the former Soviet Union and national security concerns, the dialogue did not develop into a full blown assistance program until 1992.

In 1992 at Lisbon, former Secretary of State James A. Baker announced an initiative at a conference for coordinating assistance to the NIS. The Lisbon Initiative included $25 million in safety assistance for Russia and Ukraine and really set the stage for the U.S. safety assistance program. We had begun in 1989 actual assistance in this field with the Eastern European countries after the political opening up of Central and Eastern Europe.

The 1992 and 1993 G-7 Summits were milestones in the development of the U.S. safety assistance programs. The 1992 Munich Summit for the first time focused world attention on the urgency and importance of this safety problem and resulted in an agreed action plan among the G-7 countries. This plan incorporated the three main elements of our Lisbon initiative: there were short-term immediate safety measures to address immediate operational safety enhancements for riskiest reactors, short-term risk reduction measures and regulatory assistance.

The G-7 action plan recognized that this was not just a short-term problem, however, but was also a long-term problem. The plan set in motion two steps to look at the medium and longer term issues. The first was a review of the potential for upgrading some of the newer design reactors, and the second was a study by the World Bank, the International Energy Agency and the European Bank for Reconstruction and Development of possible means to provide for alternate energy resources to replace the least safe plants.

The final pillar of that summit was the decision to create a multilateral fund to address nuclear safety. This was established in the past year at the European Bank for Reconstruction and Development. It is a donor-led account and has had substantial success in achieving the first commitment from a recipient government, in this case Bulgaria, for early closure of riskiest reactors in exchange for Western assistance.

There has been a growing consensus on how to address the problem. We have such a consensus within our own government, with the Vice President's interest, active involvement, and oversight of our interagency efforts, particularly in the case of Russia. There is also a growing multilateral consensus and institution-building to address this problem.

The group of 24 has developed mechanisms to coordinate Western bilateral safety assistance programs and to ensure lack of duplication of effort. The nuclear safety account has had success in its first project, and we expect next week to discuss possible future projects in the NIS (Newly Independent States), specifically Russia as well as Lithuania.

At the 1993 Tokyo Summit, the Heads of State began to focus on the longer term problem. They invited the World Bank and the IEA to work with the G-7 and with other lending institutions, including the EBRD and the European Investment Bank, to try to develop longer-term energy strategies on a country-by-country basis. The framework ideally would include not only those institutions but domestic credit agencies such as the Export/Import Bank and the programs of both donor and recipient countries.

In other words, it will be an effort to marshal all available resources, public and private, to try to develop strategies that would permit closure of riskiest reactors on a country-by-country basis. This is obviously a huge task. No one can guarantee its success. Energy conservation, commercial financing, market pricing, political, macro-economic, and security of supply issues abound when addressing the nuclear safety problem. But we have the makings, at least, of a multilateral effort to bring together and marshal the available resources and to try to plan on a country-by-country basis. This is a necessary step and one that is in the process of being brought together in the next month or two.

I would say just a few words about our bilateral programs. For 1991 and 1992 in Russia, Ukraine and Eastern Europe, we focused primarily on operational safety measures and programs to establish effective independent regulatory authorities in these countries. We believed that unless these steps were taken to develop an underlying safety culture, any equipment we might provide might not be used effectively.

We also counted on the countries themselves to take some of the immediate risk reduction measures to improve safety conditions in physical plant. Unfortunately, while we are encouraged by some recent developments, this did not happen so quickly, and we decided to put greater emphasis for Fiscal Year '93 programs and in our future planning for Fiscal Year '94 on risk reduction measures. As Secretary O'Leary explained, that is why the liability question has become imminent both in Russia and in Ukraine.

Secretary O'Leary mentioned the increased program for Russia for this year which we are now discussing with the Russians. We hope to resolve the liability obstacle in the very near future. I can report that I have just returned from a trip to Ukraine and Russia on this very question.

Secretary Christopher on October 25th signed an agreement that we concluded days before with Ukraine covering our bilateral assistance program which includes the liability protection provisions that we had sought.

In Russia, we are very, very close. We are agreed, I think, but for one aspect. I don't think it would be appropriate to get into the details of an ongoing negotiation, but the Vice President emphasized in his meetings with Prime Minister Chernomyrdin the importance of concluding this agreement, and Prime Minister Chernomyrdin has made a commitment to try and resolve it very, very quickly. I expect to be going back to Russia within the next 10 days to try and conclude this agreement.

I would note just a couple of things that have not been mentioned. First, I would like to emphasize the importance of the private sector in achieving overall U.S. nuclear safety objectives. We will need sustained private sector involvement to complete improvements at newer reactors. The G-7 plan, as I mentioned, does include the use of domestic credit agencies.

Second, I would note the relatively small size of the U.S. safety programs for Eastern Europe compared to the NIS. This is due to the overall relatively much smaller size of the assistance programs for Eastern Europe. It is a special challenge for us to carefully target these resources and take into account other Western bilateral assistance programs.

Mr. Chairman, in conclusion, the problem is a very large one. No assistance program can guarantee that we will not face another Chernobyl-like accident, but we are working extremely hard in a variety of fora to make our assistance as effective as possible. And we can, I think, answer the question that we continually ask ourselves, are we making a difference, in the affirmative. I think our programs have begun to make a difference, and we are looking forward to continuing and accelerating these programs.

Mr. Chairman, I would just briefly mention, as I close, the recent Ukrainian Parliament decision concerning Chernobyl. We were obviously disappointed to hear this decision. Ukraine explained it as impelled by severe electric power shortages. United States and other Western programs have not focused on Chernobyl safety improvements because we have not wanted to encourage operation of these units beyond the end of this year.

Obviously, we now have a new situation to consider. I would expect that this might be discussed at the EBRD Nuclear Safety Account meeting next week in London.

We welcome the committee's interest, support, views and the opportunity to review these programs with you today.

Thank you, Mr. Chairman.

[Testimony resumes on p. 40.]

[The prepared statement of Ms. Verville follows:]

78-866 O - 94 - 3

Senior Coordinator for Reactor Safety

and International Science and Technology Centers

Thank you, Mr. Chairman, for this opportunity to discuss U.S. efforts to improve the safety of Soviet-designed nuclear power reactors in the Newly Independent States (NIS) of the Former Soviet Union and Eastern Europe and the Department of State's role in these efforts. We have been extremely active on both the bilateral and multilateral fronts and are pleased to report considerable progress. Indeed, I have just returned from Russia and Ukraine from which I can report some new progress.

STATE DEPARTMENT ROLE

The State Department's Office of the Senior Coordinator coordinates U.S. nuclear safety efforts both within the U.S. Government and internationally. We work very closely with the State Department's Coordinators for Assistance to the Newly Independent States (NIS), Ambassador Thomas Simons, and for Eastern Europe (CEE), Ambassador Ralph Johnson, and the Agency for International Development (AID), who have been working together for the development of rational frameworks for economic restructuring and energy sector assistance generally for the NIS and CEE. These frameworks support improved energy efficiency, least-cost electricity alternatives and mobilization of public and private sector capital for alternatives to the nuclear reactors, as well as nuclear safety improvements. The Senior Coordinator also works with the technical agencies, the Department of Energy and the Nuclear Regulatory Commission, which are the lead implementing agencies for U.S. safety assistance efforts.

The Senior Coordinator has lead responsibility for establishing the overall policies and priorities to direct the U.S. nuclear safety assistance programs. These programs constitute a high foreign policy priority of the Administration. Vice President Gore provides oversight to the interagency nuclear safety assistance process led by the Senior Coordinator. In particular, the Vice President, through his co-chairmanship with Prime Minister Chernomyrdin of the U.S./Russian Committee on Energy and Space Cooperation established at the 1993 Vancouver Summit, has focused on our safety assistance efforts with Russia. On such major U.S. initiatives as those undertaken at Lisbon in 1992, at the 1992 and 1993 G-7 Summits, and under the auspices of Vice President Gore's Commission, the Office of the Senior Coordinator has played a lead role in preparing and negotiating the U.S. agendas. We also lead negotiations for the United States on nuclear safety matters in the G-7, G-24, at the Nuclear Safety Account at the European Bank for Reconstruction and Development (EBRD), and other fora, as well as with Russia, Ukraine and Central and East European countries.

I would like to highlight today our recent nuclear safety efforts, short and longer-term goals, successes and continuing obstacles. A brief review of the origins and early milestones of the U.S. nuclear safety assistance program will put these in proper context.

PROGRAM ORIGINS AND EARLY MILESTONES

Post-Chernobyl Consultations

After the Chernobyl accident in 1986, the United States began urgent consultations with the former Soviet Union on nuclear safety. Our goal at that time was modest: to begin to introduce Western ideas on the importance of safety to any nuclear power program. We were more successful than we expected. In fact, we began a two-way dialogue of mutual benefit, that expanded into limited technical cooperation and exchange of information. For several years we continued this dialogue, but at arms length, because of national security reasons.

Central and Eastern Europe

In 1989, our goals expanded. With the opening up of Central and Eastern Europe, we could no longer depend on our dialogue with the USSR to reach all of the Eastern Bloc with the Western safety message. For the most part, these nations had inherited their unsafe reactors from the Soviet Union, which provided the technical expertise and technology necessary to support operation of the reactors. When the Russians left, the United States undertook its first true nuclear safety assistance efforts at these reactors.

We sought to improve the nuclear safety situation in these countries, including better protection of the public health and safety, by improving operating conditions at the plants. Our efforts focused on (a) better operator training and provision of equipment to aid the operators in day-to-day operation of their facilities and (b) assistance to help build an effective nuclear regulatory authority through information exchanges, training, and inspector exchanges.

May, 1992 Lisbon Initiative

In 1992, changed circumstances permitted the United States to begin providing actual nuclear safety assistance to the Newly Independent States of the Former Soviet Union (NIS). Our efforts included programs parallel to those developed for Central and Eastern Europe in operational safety and regulatory assistance, and added a new feature: provision of equipment and technology to reduce the chance of another major accident at the oldest Soviet-designed reactors, the RBMKs and the VVER 440/230s. The Lisbon Initiative on Nuclear Reactor Safety was announced by former Secretary Baker at the May 1992 Lisbon Conference for Coordinating Assistance to the NIS. The Initiative included $25 million in safety assistance for Russia and Ukraine and formed the foundation of our continuing bilateral nuclear safety assistance programs to the NIS.

July, 1992 G-7 Munich Summit

The Munich Summit for the first time focused world attention on the urgency and importance of the safety problem in the NIS and Eastern Europe. The G-7 stated that each state was responsible for the safety of its nuclear power plants and that the NIS and Eastern European countries must give high priority to eliminating this danger. They noted that these efforts should be part of a market-oriented reform of energy policies encouraging commercial financing for development of the energy sector.

Several other Western countries, most notably Germany, France, Sweden and the United Kingdom, also undertook safety programs. Recognizing both the magnitude of the safety problems and the limited availability of government assistance funds, the G-7 Heads of State agreed on the importance of coordinated action to address nuclear safety.

-- They directed the newly-established Nuclear Safety Working Group of the G-24 countries to coordinate and avoid duplication of Western bilateral assistance efforts for the NIS and Eastern Europe; and

-- They agreed to implement an agreed action plan, which incorporated the three main elements from our Lisbon Initiative for short-term, immediate safety measures to address (a) immediate operational safety enhancements for the riskiest RBMK reactors and VVER 1000s; (2) risk reduction measures for RBMKs and VVER 440/230 reactors; and (3) regulatory assistance.

The G-7 action plan went beyond the Lisbon Initiative. It looked not only at the short term and immediate safety situation, but it also set the stage for actions to be taken over the longer term. The G-7 action plan recognized that the nuclear safety problem would not be solved through governmental assistance alone. Improving nuclear safety required actions not only to improve for the short term operation of the least safe reactors, but also to shut them down as soon as practicable and examine ways to find replacements for them. The action plan called for two steps towards this goal:

-- First, a review of the potential for upgrading the newer, safer Soviet designs to see if they could be brought up to acceptable safety standards.

-- Second, a study by the World Bank, the EBRD and the International Energy Agency (IEA) to study possible means to provide for alternative energy resources to replace the least safe nuclear plants.

Finally, the G-7 agreed at Munich to create a multilateral fund, directed by its donors, to address immediate operational safety and technical safety improvement measures not covered by bilateral assistance efforts.

RECENT PROGRESS

We have made substantial progress during the past year on virtually all of the above fronts, in both our bilateral and multilateral efforts.

MULTILATERAL

G-24

The G-24 Nuclear Safety Working Group (G-24 NSWG), to which the State Department leads interagency delegations, has become institutionalized with a full staff and a program of activities to ensure coordination of Western bilateral assistance efforts. It has created technical working groups through its Coordination Center in Brussels which cover country-specific (Slovakia), plant-specific (the four oldest reactors at Kozloduy, Bulgaria) and technical program areas (training). The G-24 NSWG and sub-groups, are the only multilateral fora to include recipient countries as well as donors. They meet in country and at reactor sites as appropriate to ensure full recipient participation and effective coordination of programs. In addition, several ad hoc groups have been created to address topics of common concern such as liability protection and regulatory assistance coordination. The G-24 Coordination Center in Brussels also has made good progress on an over-all data base for Western assistance programs and soon will issue a quarterly newsletter to provide the public with better information on G-24 assistance efforts.

Nuclear Safety Account (NSA) at the EBRD

The mandate of the Account, as determined by the G-7 Nuclear Safety Working Group (G-7 NSWG), is to pool donor resources for projects to provide equipment and technology for near-term risk reduction at the least safe reactors. The Account fund is managed by its donors, with assistance by the EBRD staff, separately from other EBRD accounts. The NSA has contributions of over $150 million from 13 donors and is anticipating additional contributions. Despite initial U.S. reservations about a multilateral fund, the NSA has become a multilateral success story for nuclear safety.

-- The first NSA project, for Bulgaria, was conditioned on a commitment from the government to early closure of unsafe plants. In return for grant aid from the NSA for short term upgrades at the four Kozloduy VVER 440/230 units, and an additional package of EBRD and World Bank loans which would provide Bulgaria with alternative electrical generation capacity, Bulgaria committed to closure of the four VVER 440/230 plants in the short term (by 1998) pending delivery of the above loans and grants, and timely implementation of loan project schedules.

-- In early November, the NSA staff will present the donors with a project proposal for the Ignalina plant in Lithuania. Due to the high dependence of Lithuania on these plants (80 per cent of Lithuania's electricity is provided from Ignalina) and the lack of domestic fuel resources or hard currency for purchase of fossil fuels from other countries such as Russia, early closure at Ignalina appears, at least initially, more complex than at Kozloduy in Bulgaria.

-- We also will learn in November of the NSA staff's further work plans for future projects, to include the NIS, and specifically Russia, and expect also to discuss Ukraine.

<u>1993 Tokyo Summit and G-7 NSWG</u>

At the July, 1993 G-7 Tokyo Summit, the G-7 Heads of State welcomed progress made since Munich and called for rapid implementation of donors' safety assistance, and more self-help measures by the states directly concerned. Turning to the medium and longer term, they invited the World Bank and the IEA to work with other leading institutions including the EBRD and European Investment Bank, to support the NIS and Eastern European countries in developing longer term energy strategies. Ideally, that framework will include international financial institutions, international agencies such as the IEA, domestic export credit agencies and donor countries, as well as the NIS and Eastern European countries, to marshall their resources and develop individual country energy sector strategies to identify possibilities for early closure of the riskiest plants. To encourage energy conservation and to attract commercial financing of alternative energy sources, market pricing of energy will be required.

We and our G-7 colleagues, as well as the relevant national and international institutions, recognize that the longer-term problem involves political, macro-economic and security of supply dimensions for which data may not be solid or complete. Studies completed to date by the World Bank, IEA and the EBRD also emphasize that early closure of the riskiest plants would require enormous financial resources which may be beyond the capability of all interested parties to bring together. Nevertheless, the NSWG is drawing together interested countries and institutions to explore the possibilities for coordinated measures by all concerned.

BILATERAL

Program Focus

Our FY91-92 programs for Russia, Ukraine and Eastern Europe focused primarily on operational safety measures and programs to encourage establishment in these countries of effective, independent regulatory authorities. We believed that unless steps were taken to establish an underlying safety culture in these countries, we could not be confident that risk reduction equipment we might provide would be used effectively. We also proceeded on the basis that the countries themselves would undertake immediate risk reduction measures necessary to improve plant safety. Unfortunately, they did not do so. Thus, the United States and other G-7 members concluded that the West would need to put greater emphasis on risk reduction in our assistance efforts. The U.S. FY93 program reflects this conclusion. We are continuing regulatory assistance and some operational safety programs, but have now shifted our focus to include provision of equipment for improving safety of the plants. Projects financed under the Nuclear Safety Account, as well as our bilateral programs, also reflect this new emphasis. Through recent communications and visits, we are pleased to see positive results of our assistance efforts and that the countries are now beginning to undertake some of these measures.

Country Programs

I will leave the description of assistance programs for the Department of Energy and the Nuclear Regulatory Commission. In general, we will continue our programs in Eastern Europe at about the same funding levels as in FY93 and, for Ukraine, we plan a moderate increase. I would just like to make a few observations.

First, under programs resulting from the Vancouver and Tokyo Summits, we have significantly increased funding for nuclear safety in Russia for FY 93-94. These programs will increase from approximately $12.5 million in FY92 and $15 million in FY93 to approximately $100 million in FY94. Approximately $85 million of these funds will be channeled into bilateral assistance programs -- primarily for fire safety and material conditions improvements (i.e., spare parts, maintenance training, quality assurance training) -- and regulatory assistance, which we are discussing now with Russia. The remaining approximately $15 million will be used to increase the U.S. contribution to the NSA for projects in the NIS.

Second, I want to emphasize the importance of the private sector in achieving overall U.S. nuclear safety objectives. We and the other G-7 countries recognize that over the longer term, sustained private sector involvement will be crucial to complete nuclear safety improvements at newer reactors. Domestic credit agencies such as the Export-Import Bank (EXIM) can play a critical role in those countries which meet EXIM lending criteria. For example, EXIM is considering a loan to a U.S. company, Westinghouse, which was selected by the Czech Government to undertake important upgrade work at the VVER 1000 Temelin nuclear reactor. This kind of financing can be important to maintain momentum for nuclear safety improvements for the future, and the G-7 NSWG will include the potential for domestic export credit agency financing in the framework it is developing to marshall all potentially available resources for energy sector plans to permit early closure of the riskiest reactors.

Third, I would note the relatively small size of the U.S. safety assistance programs for Eastern Europe compared to our NIS program. Due to the relatively much smaller size of the overall U.S. assistance program for Eastern Europe, our nuclear safety program for all East European countries combined, i.e., the Czech Republic, Slovakia, Bulgaria, Hungary, plus Lithuania, is approximately $5 million for FY94. We are working hard to target these resources as carefully as possible, taking into account other Western bilateral assistance programs as well as NSA and other multilateral assistance.

Gore-Chernomyrdin Commission

U.S. nuclear reactor safety assistance efforts with Russia are a central part of the Vice President's agenda with Prime Minister Chernomyrdin. The Vice President's use of this high level forum to emphasize U.S. concerns about nuclear safety in Russia has given real political impetus to our programs.

-- At the first Commmission meeting, the Vice President personally emphasized the need to conclude our bilateral nuclear safety agreement to provide, among other things, liability protection, for the United States and its contractors providing safety assistance at Russian reactors. Prime Minister Chernomyrdin's personal interest has resulted in a Russian Government commitment to its early conclusion.

-- The Vice President and Prime Minister also agreed to a joint study of possible alternate sources of energy in Russia in order to explore possibilities for early closure of riskiest plants. This study is being carried out through the Agency for International Development in close cooperation with the World Bank and other institutions, drawing on the work already done by them and seeking to reach conclusions. Again looking to the medium and longer term, this cooperative study acknowledges the reality that reactor safety must be viewed in the overall energy context.

OBSTACLES

The major obstacle to implementation of U.S. nuclear safety assistance programs has been lack of protection for the United States and its contractors from potential liability for damages or injury caused by a nuclear incident at a plant at which the United States and its contractors are providing safety assistance. We are concerned about potential liability both within and outside recipient countries. The problem is particularly acute due to the absence of domestic nuclear liability laws in Russia, Ukraine and Eastern European countries, that are in any way similar to those in the United States and other Western countries. Nor have the recipient countries yet joined the Vienna Convention on Third Party Nuclear Liability.

In these circumstances, the highest U.S. priority has been to complete negotiations on bilateral nuclear safety assistance agreements with Russia and Ukraine, where our program plans make liability protection an urgent matter. I am pleased to report that we concluded negotiations on an agreement with Ukraine last week containing liability protection. Secretary Christopher signed the Agreement on October 25. Negotiations in Russia October 14-15 resulted in agreement on all but one aspect of the problem. Senior officials on the Russian side told us that they see no insurmountable obstacle to a resolution of this issue in the near future. While nothing is agreed until everything is agreed, we are optimistic that this Agreement can be concluded soon, so that we can avoid disruption in the U.S. program.

CONCLUSION

In conclusion, Mr. Chairman, we do not underestimate the magnitude of the problem of unsafe reactors or the financial, political and economic obstacles to resolution of these problems. No assistance program can guarantee that the world will not face another Chernobyl-like disaster. What I, along with Secretary O'Leary and Chairman Selin, can assure you is that we are working very hard in a variety of fora to make our assistance and programs as effective as possible. We are coordinating and marshaling resources around the world to apply to this issue. The program is receiving high-level attention and priority under the Vice President's oversight.

The bottom line question, which we continually ask ourselves, is: Are our efforts making a difference? Have we made the riskiest plants safer or reduced the risk of a serious nuclear incident? We think we are making a difference and that we are reducing this risk. Our initial efforts beginning in 1991 on operational safety and regulatory measures have begun to show results. Through such programs as exchanges and seminars, as well as efforts to establish emergency operating procedures under Secretary O'Leary's programs, we have begun to improve the operational safety awareness and performance at least in some of these countries. Chairman Selin's efforts also have accomplished a great deal with relatively low expenditures to help create and give proper status to independent, effective regulatory authorities, again, in at least some of these countries. However, even this job is far from done. For example, the highly competent and energetic Ukrainian NRC Chairman equivalent operates on a shoe string and without even the most basic office facilities. In Russia, the regulator's role still needs further emphasis within the government.

The United States and other G-7 countries, particularly through the NSA, are now concentrating major efforts on risk reduction measures such as providing fire protection and detection equipment, diesel generators and other equipment to improve physical plant safety. The liability obstacle has now been addressed in Ukraine and, we hope, is very close to resolution in Russia so that we can proceed without disruption in our programs. We also have the aggressive agenda for the Gore Chernomyrdin Commission and are working hard through that mechanism and in the G-7 NSWG to complete overall energy sector assessments and marshal available resources.

In short, we believe we are making a difference and hope to continue and even expedite the progress we have made to date. We cannot report that we have solved the problem, or even that it will be fully solved in the near term. Much remains to be done in all areas - risk reduction, operational safety and regulatory reform - as well as on longer term planning. But that strengthens our resolve rather than discourages us. We cannot afford to take any other view, or not to continue the program, would seem to us unthinkable.

I would like in closing to briefly address the recent Ukrainian Parliament's decision to lift its prior ban on completing construction of newer model Ukrainian nuclear power plants as well as its previous order to close all Chernobyl units by the end of this year. Ukraine explained its decision as impelled by severe electric power shortages. The Ukrainian announcement raises new issues not only for the U.S. bilateral assistance program, but also for the wider multilateral work we are working with in the G-7 context.

The U.S., and most other western, programs have not focused on

Chernobyl safety improvements, because we have not wanted to encourage operation of the units beyond the end of this year. Now that Ukraine has made clear that it cannot keep to the 1993 shut down schedule, we may have to consider the safety implications of this decision. We hope that this will only be a short term issue. We are urging Ukraine to cooperate with the on-going G-7 work to develop country energy sector strategies that will provide the assessment of the potential for energy alternatives to permit early closure of the Chernobyl units. For example, some international financing may be available for alternate sources if Ukraine is able to meet a minimum standard of credit-worthiness, which combined with possible commercial involvement in completing newer, safer plants and energy policy reforms by Ukraine, could permit early closure of the Chernobyl units. Mr. Chairman, these remarks are only preliminary since this is a new development, but we are placing high priority on consideration of the U.S. and international response to Ukraine's decision. I would expect this to be discussed at the EBRD meeting next week in London.

Mr. Chairman, we welcome the Committee's interest, support and views. Again, I appreciate this opportunity to review our programs and goals with you and look forward to our continuing close cooperation to address these important issues.

Mr. SHARP. Thank you, Ms. Verville.

Chairman Selin, we are very pleased to hear from you now.

STATEMENT OF HON. IVAN SELIN

Mr. SELIN. Thank you very much, Mr. Chairman.

As I am sure you know, I have visited all four types of the common Soviet-designed reactors and a couple of the other types in Russia and the Ukraine and the six other countries that operate the Soviet-designed plants, and following the guidance of your cover letter, I thought it might be useful to just describe some of the features of the Chernobyl type RBMK and of the oldest type of pressurized water reactor, the VVER 440 Model 230 that, in fact, causes so much concern.

The Chernobyl type design was intended as a military production reactor which could also produce electricity. In its original design, it had little of the attention paid to safety that we expect in the West. There have been improvements in each subsequent generation, but we still consider the overall design to be flawed. Some of these design problems are unique to the RBMKs, some of them are common to other Soviet designs, but none of them occur in the modern Western reactors.

To summarize, the major design deficiencies of the RBMKs are that, first, they are characterized by a general lack of separation and redundancy in critical electrical and fluid systems, so their safety systems show susceptibility to failures that have a common cause. In spite of the fact that they may have three diesels and three sets of pumps, the same accident could put all three trains out at the same time.

The second is that none of the RBMK's exhibit adequate levels of fire protection, the emergency core cooling systems are limited in effectiveness to pipe breaks in only a small number of locations. They have no containment system. In the RBMK there is a confinement system, a system that could keep radiation from going to the atmosphere, but this confinement system can't withstand high pressure and, furthermore, the break of more than one or two of the 1,661 tubes out of which the core is constituted would overcome this confinement system.

Moving to the VVER 440/230, this was the first production model of the pressurized water reactor concept which is the way most Western reactors are built. It was designed in such a manner that it does not measure up to Western safety requirements. There are some inherently positive features in the design. It has a large water inventory and it has low power density. The fuel is spread over a large volume, and so not so likely to cause a problem but, nevertheless, there are some major design deficiencies in this model as well. It doesn't have a containment. There is also a confinement, but it would only work with the smallest possible breaks in the primary coolant system. There is no substantial emergency core cooling system if the basic cooling system fails. It has the same lack of separation and redundancy in emergency systems that I cited for the RBMKs. The systems are subject to common cause failures, and it also shares with the RBMK an inadequate built-in protection against fire.

Again, I would like to emphasize that these faults distinguish the 440/230s from the Western reactor designs. There is consensus among Western nations that the remaining 15 RBMKs in the former Soviet Union and the ten 440/230s spread throughout the former Soviet Union and Central and Eastern Europe should not be operated any longer than necessary, but we do believe that short-term risk reduction measures are justified to prevent an accident in the short-term, and also as a way to build up the confidence that we think is necessary to give us some leverage in engaging in the types of discussions that Congressman Markey cited.

Secretary O'Leary already cited the September Gore/ Chernomyrdin meetings, but from a regulatory viewpoint, I would like to stress one outcome, namely that Prime Minister Chernomyrdin committed Russia to examine carefully the need for a stronger independent regulatory body. The NRC will be working with Russia to improve its regulatory structure. There are serious weaknesses and the Russian regulatory body will have to assist them in developing the legislative, the regulatory and the liability framework for it to be an effective regulatory organization. There is broad agreement of both West and East that a strong independent regulator is a prerequisite to operational nuclear safety.

The steps that are called for include the development and acceptance of a legal basis for a strong and independent regulator. There is now no Russian law setting up the Russian regulator. It exists entirely by edict and it is part of the Ministry of Atomic Energy. We will work towards the provision of adequate resources, both material and personnel, to fund and staff an organization with the ability to monitor plant safety and operations. It should have the authority to intervene in operations to insist on safety and, when and if necessary, to shut down a plant in face of danger to the public. The Russian Federation should adopt internationally agreed safety principles including public accountability through openness about reporting incidents and accidents which have safety implications and giving the public a voice in reactor licensing.

One possible outcome of our efforts was that on September 16 President Yeltsin issued a decree significantly expanding the authority of the Russian nuclear regulator, Gosatomnadzor or GAN. GAN has now been officially declared the lead nuclear safety oversight organization in Russia. Its sphere of influence has been expanded to cover all nuclear facilities, but we are still very concerned about whether they have adequate budget and other resources to do their job as well as a solid legal basis and the day-to-day authority to carry out their work.

In summary, Mr. Chairman, the countries operating Soviet-designed reactors have already made significant progress in short-term risk reduction, although a lot more is needed. Substantial assistance will soon be needed from the United States and others to further reduce the risks of a serious nuclear plant accident in these countries.

The progress on the ground masks the continuing fundamental problems with the RBMKs and the 440/230s. We see some short-term risk reduction, but the long-term problem is almost as serious as ever. Obviously, ultimate responsibility for safety lies with the states of the former Soviet Union and the countries of Central and

Eastern Europe. We can help and we must help but we cannot supplant their responsibility.

The difficult step will be to convince them that even after the short-term fixes in these riskier reactors that these reactors still need to be replaced and they need to be replaced rather soon.

The interest and concern of the international community, as Ms. Verville cited, have been very encouraging and, in short, we now have a historic opportunity. I sincerely hope it will not be missed. Your witnesses and other members of the government are working as hard as we can to make sure that this opportunity is not missed.

That concludes my oral statement, Mr. Chairman.

[Testimony resumes on p. 59.]

[The prepared statement of Mr. Selin follows:]

STATEMENT OF

IVAN SELIN

CHAIRMAN

U.S. NUCLEAR REGULATORY COMMISSION

Mr. Chairman, Members of the Subcommittee, I am pleased to be given the opportunity to address the important issue of improving the safety of Soviet-designed nuclear reactors in Central and Eastern Europe and the former Soviet Union (FSU), particularly the Nuclear Regulatory Commission's role in this effort.

I have just returned from a trip to Central and Eastern Europe (CEE), during which I served as a member of the United States delegation at the General Conference of the International Atomic Energy Agency (IAEA) in Vienna. All eyes and ears seemed tuned to what had transpired only three weeks earlier in Washington when Russian Prime Minister Chernomyrdin met with Vice President Gore to discuss, among other things, improving nuclear reactor safety in Russia.

This was my third trip to this area, having visited a number of the same countries and sites both in 1991 and 1992, and I can report progress: on the ground, in the material condition of many of the plants that were so seriously flawed as late as just two years ago; in the sensitivity of the leaderships of these countries to the safety concerns of the West; and, most importantly from my point of view, the growth (however slow) of strong and independent regulatory bodies that will eventually be capable of exercising the same kind of authority over safe operations that the NRC does in the United States.

The NRC has developed into a unique nuclear regulatory body, one, I might add, after which a number of other similar bodies have patterned themselves in other countries. We have the authority to take action to close down plants we find to be operating in an unsafe manner in order to protect the public health and safety. NRC regulations are used as models for many of the safety codes and guides of international bodies such as the IAEA. NRC experts are called on to participate in working groups to develop international consensus on new safety principles. We operate an independent, confirmatory research capability that enables us to develop our own perspective on problems from severe accidents, to the quality of the nuts and bolts used in our reactors; from the relative risks involved in operating these plants, to the ultimate question of how long these plants should be allowed to operate. We do not need to depend on vendors or utility operators for these judgments.

The countries of Central and Eastern Europe, and particularly the former Soviet Union, do not have this tradition of regulation. In Russia, for example, the nuclear regulatory body does not operate under a national nuclear law such as that which set up the NRC, but has derived its authority as a part of the Ministry of Atomic Energy. One of the goals of the Gore/Chernomyrdin Commission (GCC) is to improve both the legal and political stature of the regulator in Russia, to give it a place at the table so to speak, so that it commands the respect of both the Ministry and the utilities operating the nuclear power plants. The shutdown of unsafe plants in these countries will ultimately depend upon the strength and independence of their regulators.

With the level of effort we in the U.S. Government have exerted over the last two years, the short-term phase of western assistance programs to the former Soviet Union and Central and Eastern Europe is beginning to show signs of real success. Even one of our usual critics, Viktor Mikhailov (head of Russia's

Ministry of Atomic Energy) recognizes this progress. He told the European Parliament on September 22 that bilateral nuclear relations with the U.S., while more recent than those with the European Community, had produced "better results."

As we settle the liability question in Russia and Ukraine and in the Baltic States, near term improvements will buy time to improve overall energy efficiency and should make possible the closure of the least safe plants. The NRC's role in this assistance effort is unique, since it is manpower-, rather than equipment-, intensive. NRC programs involve inspector training and institution building, rather than hardware backfits, and are thus quite economical to implement. Government and regulatory leaders from many of these countries personally expressed their appreciation to me for this assistance, and I was able to witness some of the effects in my visits to various sites.

Paradoxically, western success in securing near term improvements in safety, will require us to confront earlier the basic problem of closing down the least-safe reactors as soon as practicable. The program of western nuclear safety assistance developed at the 1992 G-7 Munich Summit was not intended to extend the life of these reactors indefinitely. However, it is admittedly difficult to draw a fine, bright line between near-term safety upgrades and improvements which could encourage an operator to think in terms of long-term life extension. The task of identifying and financing replacement power for these facilities is a very difficult one, potentially involving many billions of dollars of capital which neither we nor the affected countries have readily available. Relevant international financial institutions (the World Bank, European Bank for Reconstruction and Development (EBRD), and International Energy Agency (IEA)) have prepared useful studies of these long-term issues, but it will be difficult to engage the Russians (and other nations operating Soviet-era plants) to take the hard decisions needed to close the worst plants.

Measuring success in our assistance program will, therefore, be difficult, and, much as we would like to see the many unsafe reactors shut down immediately, it is not likely to happen any time soon. Accordingly, the U.S. has tried to direct attention to business-like ways of finding replacement sources of electric power, without which the Russians will not begin shutting down even their oldest and riskiest plants, and this is a key element of the Administration's follow up to the Gore/Chernomyrdin meetings. Russian energy price decontrol will be crucial to encourage conservation and to facilitate commercial funding and repayment of loans for safety enhancements and replacement power.

One sign that <u>will</u> indicate progress will be when the regulatory bodies in the NIS and CEE have the authority and the will to shut down nuclear power plants for safety violations. A significant portion of the NRC effort to implement the Gore/Chernomyrdin Commission mandates will be to strengthen the Russian nuclear regulator -- Gosatomnadzor (GAN).

During the September meetings, the Vice President was able to elicit a commitment from the Prime Minister to a much strengthened nuclear regulatory body, a goal towards which the NRC has been working since 1988. As one possible outcome of our efforts, on October 16 President Yeltsin issued a decree significantly expanding the authority of GAN. It has now been officially declared the lead nuclear safety oversight organization in Russia and its sphere of influence has been expanded to cover all nuclear facilities. Its funding will derive from a separate line item in their Federal budget. While we see this as forward progress, we have yet to learn the size of the separate budget for GAN, but we expect to press the Russian Government at the highest levels to assure full and adequate funding for all GAN's responsibilities, including salaries for its inspectors. In addition, GAN will sit at the table with Minatom during all future Gore/Chernomyrdin negotiations on nuclear safety matters.

NRC is in a position to deal with Russian regulators thanks to a cooperative U.S./Soviet technological effort that has been underway since 1988. The recent meetings between Vice President Gore and Russian Prime Minister Chernomyrdin are a continuation of this relationship, and they have taken the relationship a giant step into the future. To illustrate this, perhaps a brief review of recent summit activity on nuclear safety would be useful.

The Lisbon Initiative

In May 1992, at the Lisbon Conference on Assistance to New Independent States, the U.S. announced several initiatives. Prominent among these was a nuclear safety initiative to spend $25 million on:

- Cooperative activities with Russia and Ukraine, building on prior work done by the Department of Energy and the Nuclear Regulatory Commission under a 1988 Memorandum of Understanding;
- Establishment of two regional training centers, one in Russia and the other in Ukraine;
- Immediate operational safety enhancements for VVER 1000s and VVER 440/213s;
- Risk reduction measures for RBMKs and VVER 440/230s; and,
- Regulatory assistance in developing consistent and effective safety standards and procedures, as well as training in their use.

This initiative, along with a broader nuclear safety initiative launched at the G-7 Summit Meeting in Munich in July 1992, marked a significant new departure in efforts to improve safety in Central and Eastern Europe by involving the U.S. Government at the level of head of state.

The 1992 and 1993 G-7 Summits

About a month after the Lisbon Conference, the Group of Seven (G-7) heads of state, meeting at Munich, recommended a Program of Action that had been prepared by a Nuclear Safety Working Group in May 1992. The Working Group found a growing international consensus that the remaining 15 RBMKs in the former Soviet Union and the 10 VVER 440/230s in the former Soviet Union and Central and Eastern Europe should not be operated any longer than necessary, although short-term risk reduction measures would be justified, and that the better-designed plants (VVER 1000s and 440/213s) should be upgraded to achieve an acceptable level of safety. In addition, reactors of the newer design under construction could be completed if economic conditions warranted, and funded through commercial loans. (There are a total of 17 VVER 1000s in Russia and Ukraine, two each in Bulgaria and the Czech Republic; there are four of the smaller VVER 440/213s in the Czech Republic, two in Slovakia, and four in Hungary.)

While western countries had no authority or leverage to demand that the worst plants in Russia be shut down, the Summit recommendation was intended to promote agreement in principle with the republics of the former Soviet Union to limit the remaining lifetime of the less safe plants. In addition, the program was to create the basis for longer-term safety improvements by examining the scope for replacing less safe plants by the development of alternative energy sources and more efficient use of energy, and the potential for upgrading plants of more recent design. Further, the Summit members promised to pursue the early completion of a nuclear safety convention.

I should emphasize one key principle of our joint assistance efforts, namely, that improving the regulatory structure in the countries of concern is an essential element in achieving nuclear safety. But equally important, in seeking upgrades for nuclear plants and improving operational safety, is the achievement of a fundamentally new attitude towards energy economics on the part of officials in Russia and its neighbors. It will be necessary to assist them in developing the legislative, regulatory, and liability frameworks for an effective, independent regulatory infrastructure, and lead them to a market based energy economy to generate revenues adequate to maintain their nuclear power plants.

The Gore/Chernomyrdin Commission

At the Vancouver summit between Presidents Clinton and Yeltsin in May of this year, the U.S. pledged $15 million to assist Russia in nuclear safety. There will be an additional $15 million for Ukraine. Of perhaps even greater importance, however, the two Presidents asked the Vice President of the United States and the Prime Minister of the Russian Federation to establish and chair a Joint Commission on Energy and Space. The Commission was founded to establish a dialogue between the two governments at the political level for expanding cooperation in energy, nuclear safety, and space technology as well as in several other areas, and to serve as a forum for jointly resolving practical problems in this expanding relationship.

U.S. objectives in providing nuclear reactor safety assistance are to help Russia:

- reduce the near-term risk of operating the less safe Russian reactors; and

- reinforce the authority of an independent Russian regulatory regime; while,

- rationalizing the Russian energy economy through market pricing, conservation, and use of alternative energy sources.

In support of these objectives, the Gore/Chernomyrdin Commission had as its goal the communication of Western safety concerns and policy perspectives to the highest levels of the Russian government. Although we would like to see the VVER 440/230 and RBMK plants closed down, it appears for various reasons that the Russians and others are going to operate them for a while, and we cannot ignore this fact. Therefore, the U.S. has made significant investments in short-term risk reduction which we hope will have the additional benefit of improving national confidence such that we will have laid a solid basis for discussions concerning their decision to continue operation of their less safe reactors in the future.

The two key messages the Vice President wished to give Prime Minister Chernomyrdin was to express U.S. and Western concern about the continued operation of the least-safe reactors, and to press for a strengthened nuclear regulatory authority. NRC helped with preparations for the meetings, which emphasized U.S. policies:

to encourage the Russians to introduce risk reduction measures in their least safe plants;

to give greater autonomy and authority to the regulatory body Gosatomnadzor (GAN);

to improve operational training through the use of simulators;

to develop emergency operating procedures;

to complete arrangements for liability protection to enable U.S. industry to provide safety assistance, and,

to gradually replace their obsolete Chernobyl-type RBMK reactors with modern PWR's equipped with containment and augmented safety systems on the model of most western reactors.

The NRC also took the lead in arranging a visit to the St. Lucie Nuclear Power Plant in Florida for Prime Minister Chernomyrdin. I was privileged to host the visit, and we held frank and productive discussions with Prime Minister Chernomyrdin and his top advisers, both in Washington and Florida. Prime Minister Chernomyrdin indicated that he was greatly impressed with his St. Lucie visit and said he believed U.S. cooperation could provide a valuable contribution to advancing nuclear safety in Russia. The Vice President raised the importance of an independent nuclear regulator. The Russians agree that the ultimate responsibility for nuclear safety resides with the operators of the power plants, not with the regulators. Nevertheless an independent, legally constituted, well funded safety regulator can assure that the operators achieve the proper degree of vigilance and devote proper attention to safety. The Vice President and Prime Minister suggested that this issue should also be followed up in further Commission activities.

Prime Minister Chernomyrdin committed Russia to examine carefully ways to strengthen the independent regulatory body. The U.S. will work with Russia to improve its regulatory structure, based on commonly agreed criteria for such a body. NRC is the appropriate U.S. agency to take the lead in assisting Russia in regulatory enhancement. The insight gained by NRC through our Joint Coordinating Committee on Civilian Nuclear Reactor Safety (JCCCNRS) which has been in operation since 1988, as well as certain internationally agreed principles of regulation and reactor safety, provides a solid foundation for the conceptualization and development of a set of recommendations.

During the Gore/Chernomyrdin session the two sides agreed to form a subcommittee on nuclear safety. Secretary O'Leary and I are co-chairs for the U.S.; the Atomic Energy Minister and the Nuclear Regulator are the Russian co-chairs.

The key elements of nuclear regulation should include (but not be limited to):

- Development and acceptance of a legal basis for a strong and independent regulator;

- Provision of adequate resources, both material and personnel, to fund and staff an organization with the ability to monitor plant safety and operations;

- The authority to intervene in operations to insist on safety and, when and if necessary, to shut down a plant in the face of danger to the public;

- The adoption of internationally agreed safety principles; and,

- Public accountability through openness about reporting incidents and accidents at plants which have safety implications, and a public voice in reactor licensing.

In fact, many of NRC's assistance activities focus on these key points.

Prime Minister Chernomyrdin also expressed his willingness to sign agreements on both 1) Radiation Health Effects with the United States, and 2) Liability Protection for western companies doing business in Russia. The Departments of State and Energy have the lead in the development of the Health Effects Agreement; the State Department in its negotiation, and the Department of Energy in its implementation. The NRC will participate and assist in the negotiations, and was represented on the negotiating team in Moscow in mid-September, where such an agreement was initialed.

Russia has neither acceded to existing international liability regimes nor enacted domestic legislation related to nuclear liability. The U.S. began negotiating a bilateral agreement with Russia in early 1993 to provide adequate legal liability protection for U.S. companies providing nuclear safety assistance to Russia. The proposed agreement is intended to ensure that the Russian Federation is held legally responsible for the operation of its nuclear power plants, including responsibility for the consequences of any incidents that may occur, as is the case in most western countries.

Prior to the Commission meeting, although Russia had expressed its intent to conclude arrangements, indications were that it could take several years to complete this process. During the Commission meetings in Washington, Prime Minister Chernomyrdin and Atomic Energy Minister Mikhailov indicated their determination to resolve the liability question as soon as possible. In addition to a bilateral agreement with the U.S. to resolve the near-term assistance problem, the Russians need to move forward with domestic legislation to deal with the liability issue in a more regular way.

Secretary O'Leary followed up the Washington discussions in meetings with the Russian leadership in Moscow late last month. (My own follow up meetings were preempted by the battle for the Russian White House on October 4.) The U.S. has effectively engaged the Russian political leadership in a fruitful dialogue on nuclear safety in the context of overall economic and energy development. This is a major achievement, and one without which it might have been impossible to have a significant impact on Russian nuclear safety plans for the future. While the details of an expanded joint U.S. - Russian nuclear safety program are only now under negotiation, and are expected to be finalized during a visit by the Vice President to Moscow in December, I believe we have turned a corner into a new, more productive path in which nuclear safety improvements will be taken in the context of an overall energy strategy that will emphasize market forces, market pricing of energy, and viable alternatives to Russia's least safe nuclear power plants.

Evaluation of Status Today

While the U.S. is placing high national priority on assistance, particularly to Russia, improvements are slower than we hoped. There has been significant progress in short-term risk reduction at the plant level. Nevertheless, while they praise the U.S. programs, Russian officials continue to complain that assistance is not getting through to the plant and operator level. We will be discussing this matter with our Russian counterparts under the new structures for cooperation set up by the Gore/Chernomyrdin Commission. Too much of the western funds continue to be spent in donor countries. Economic conditions and decisions in many of the recipient countries today do not favor improved nuclear safety.

In addition, because of their need for power, Russia and Ukraine have shown an interest in resuming construction on several nuclear power plants. Last week the Ukrainian parliament lifted a moratorium on nuclear plant construction, which will enable them to finish some needed new capacity in VVER 1000's, but at the same time they decided to continue operation of Chernobyl. Most worrisome is the attitude displayed in Russia that there are no fundamental problems with the RBMKs and therefore that their lifetime can be extended safely to a full 40 years.

Of particular urgency is the continuing need for regulatory authorities to be strengthened in Russia and Ukraine. Ukraine is slowly developing a government infrastructure for dealing with nuclear safety and is facing energy needs that they believe require keeping the Chernobyl plant open. We at NRC are responding to what the regulators in these countries have asked for, and we need to stay the course. We have also established project agreements with the Czechs, Slovaks, Hungary, Bulgaria, and Lithuania, to help them improve regulation. But our task will require a long term effort and continued funding, both bilateral and multilateral, public and private.

With regard to Eastern European regulators, I am pleased to report that in a recent meeting in Vienna on September 30, all regulators from countries owning VVER 440 reactors agreed to collaborate and meet on a regular basis jointly to address problems in operating these Russian designed reactors. The NRC and the German regulators have been asked to participate in these meetings as associate members of the group. The NRC can work through this group when we are providing common assistance. Most importantly, this group can share experience in operations, training, maintenance, and other areas of regulatory interest for VVER 440 reactors.

Not only does the U.S. need to maintain current levels of funding for short term programs, we also must encourage market pricing of energy to increase financing, or the conditions for financing, of long-term energy options. Finally the U.S. needs to support multilateral funding efforts through international development institutions such as the World Bank and the EBRD.

Mr. Chairman, this completes my statement. I will be pleased to answer any questions that you and the Subcommittee may have.

ATTACHMENT

THE HISTORY

During the past seven years the NRC has devoted extensive efforts to nuclear safety cooperation with, and assistance to, countries with Soviet-designed reactors in the former Soviet Union (FSU) and Central/Eastern Europe (CEE). These efforts have been in three phases: 1) the early post-Chernobyl phase (1986-1988), 2) development of cooperative activities under a Joint Coordinating Committee on Civilian Nuclear Reactor Safety, or JCCCNRS (beginning in 1988), and 3) technical assistance under the Lisbon Initiative on Nuclear Safety from May 1992 until the present.

Cooperation with, and assistance to, the former Soviet Union

Preliminary interchange on Russian nuclear reactor design and operation characterized the first phase, in which the U.S. tried to ferret out the causes and consequences of the Chernobyl nuclear accident of April 26, 1986. The Soviets were naturally suspicious, as hundreds of foreigners sought access to what had been a highly secret nuclear establishment. The major accomplishment of this stage was to gain the Soviets' trust and to help them realize that the West had much to teach them about nuclear safety, and perhaps also had something to learn.

The second, more substantive phase began with the signing of a Memorandum of Cooperation on the second anniversary of the Chernobyl accident, April 26, 1988. Interactions took the form of ten (later to grow to twelve) working groups on technical aspects of civilian nuclear power reactor safety, under the supervision of a Joint Coordinating Committee on Civilian Nuclear Reactor Safety (JCCCNRS). In the following years about fifty scheduled conferences on nuclear safety were held alternately in USSR/FSU and the U.S.

This series of exchanges, stretching over five years, has been of considerable value to the U.S. and the NRC as well as to the nuclear establishment in Russia and Ukraine, both their regulators and their nuclear power planners. Some of the information that has been exchanged has been translated into activities which have benefited reactor safety worldwide.

- **Information Exchange**: Through 1992, about 50 technical meetings in the FSU and the U.S. led to a deeper understanding by the FSU of the technical, legal and organizational approaches to safety employed in the West and had a positive influence on Russian and Ukrainian safety culture, making them better able to help themselves improve safety.

- **Inspector Exchanges**: An unprecedented series of exchanges of inspection personnel for extended periods at nuclear facilities benefitted regulatory authorities in the USSR, broadened U.S. understanding of the Soviet regulatory infrastructure, and led to efforts to change some Soviet policies and practices.

- **Severe accident research**. The Russian Kurchatov Institute, working with NRC but using Russian facilities whose technical equivalents are not available in the U.S., is contributing to the NRC mission. As a significant milestone in this work, Kurchatov's safety research director spent several months at NRC in 1990 to learn how safety research is done in the U.S. so this knowledge could be applied in Russia. In addition, the NRC provided to Kurchatov U.S. codes for analysis of severe reactor accidents.

- **Involvement in Western technical organizations**: In exchange for access to the accident codes, Russia has become a member of the International Code Assessment Program, an association of Western countries interested in simulating reactor accidents.

- **Russian annealing technology**: Through the exchanges the U.S. has learned much about Soviet annealing technology. NRC learned that the Soviets had annealed (heat-treated) reactor pressure vessels that had become embrittled by radiation and they had obtained almost complete ductile recovery. Because of similar problems of embrittlement being encountered in U.S. reactors, the NRC is interested in this technology, and some utilities with older plants are also investigating its promise.

In the course of these programs, involving a large number of technical visits, the U.S. also learned about the poor condition of most nuclear power stations in the former Soviet Union. Therefore, when the Soviet Union dissolved at the end of 1991, the NRC was in the forefront of U.S. efforts to help improve safety and regulation in Russia and Ukraine, the two resulting republics that had large nuclear infrastructures. The resultant programs of assistance drew on the cooperative activities of the previous years, serving as the basis for a U.S.-led international effort to improve reactor safety.

The May 1992 Lisbon Initiative began the third phase of U.S. nuclear safety cooperation with the former Soviet Union. This phase emphasized direct assistance, building upon the program of technical cooperation. The Secretary of State announced a $25 million U.S. nuclear safety assistance program for the FSU which included a program of nuclear safety assistance in Russia and Ukraine. Shortly after the announcement, NRC met with Russian and Ukrainian regulators to solicit their ideas for enhancing their regulatory capabilities. The programs developed for regulatory assistance are mainly oriented toward training in licensing, inspection, research, and emergency response; but they also include provision of analytical equipment.

Program training is conducted during group visits of Russian and Ukrainian regulators at NRC headquarters, regional offices, National Laboratories, and the NRC Technical Training Center (TTC) in Chattanooga, Tennessee, or when NRC staff provide briefings or seminars in Russia or Ukraine. Of some 60 visits planned, over half were underway or completed by the end of FY 1993, involving about 100 experts, trained for an average of three weeks, with some staying for as long as three months. Those assigned to the U.S. program report the training is effective because it is specific and technical; and, because it is outside their homeland, it offers a different cultural perspective as well.

Cooperation with, and assistance to, Central and Eastern Europe

Beginning in April 1989, with the signing of an NRC bilateral information exchange arrangement with the Czechoslovak Atomic Energy Commission, NRC's cooperation with Central and Eastern Europe (CEE) took on some of the intensity of the Soviet program. In September 1990, NRC signed an information exchange arrangement with the Hungarian Atomic Energy Commission, which covered topics similar to those in the arrangement with the Czechs. In late 1990, NRC sought funding for assistance to Central and Eastern European countries from the Agency for International Development (A.I.D.), and in September 1991 concluded an agreement with that Agency whereby NRC would be provided $575,000 in funds during fiscal year 1991 in support of nuclear safety assistance to the Czech and Slovak Federal Republic (CSFR).

On January 31, 1992, this agreement was amended to provide an additional $150,000 for technical assistance to Bulgaria, bringing the total for fiscal year 1991 for the three countries to $725,000, for such tasks as nuclear safety orientation training, an NRC Fellowship Program for on the job training alongside NRC staff, training at the NRC Technical Training Center, and membership fees in the "Severe Accident and Risk and Accident Management Programs" and the "International Code Assessment Program."

In 1992, Lithuania and the Slovak Republic were added to the Central/Eastern European countries receiving regulatory assistance from the NRC. NRC asked for and received an additional $900,000 for fiscal year 1992 from A.I.D. in support of continued and expanded nuclear safety assistance to the Czech Republic, the Slovak Republic, Hungary, Bulgaria and Lithuania.

By the end of July 1993, 34 regulatory personnel from Bulgaria, the Czech Republic, Slovakia, and Hungary will have been trained by NRC for an average of 1-1/2 weeks; two assignees, one from Hungary and one from the Czech Republic, also participated in NRC's Fellowship Program for six months; the Czech representative returned to become head of the nuclear regulatory organization in the newly formed Czech Republic. The technical exchanges and training conducted under NRC's cooperative programs are helping to encourage an enhanced safety culture in Central and Eastern Europe.

THE PROBLEM

The safety problems of Soviet-designed and built reactors, especially the 11 Chernobyl-style RBMK's (4 at Kursk, 4 at Sosnovy Bor, and 3 at Smolensk) and 4 VVER 440/230 models in Russia, include poor plant design (neither model has containment), siting, construction, quality assurance, operations, maintenance and regulation.

Following the Chernobyl accident in 1986, the U.S. and others analyzed the accident and the type of reactor involved, the RBMK, and subsequently analyzed the other main type of Soviet-designed reactor, the VVER, as well as the Soviet nuclear power and nuclear safety programs. These analyses found a generally poor safety culture resulting in operational safety concerns throughout the former Soviet system. In addition, significant design deficiencies were found in most of the models which would not meet western safety standards.

The RBMK reactor design, for example, was developed as a military production reactor which could also produce electricity, with little appreciation for nuclear safety as we understand it in the West. While there have been improvements in each subsequent generation, the overall design remains seriously flawed. The major design deficiencies of the RBMKs are:

- Fundamentally unstable reactor design and poorly designed systems for shutdown of the reactor.
- Lack of separation and redundancy in critical electrical and fluid systems and general susceptibility of systems important to safety to common cause failures.
- Little or no fire protection.

- No containment system and a confinement system which does not enclose the entire primary coolant system and with limited pressure retention capability.

Concern over the operation of these reactors remains substantial. All of the plants have serious shortcomings in operational safety, material condition and in design, although some progress is being made on the ground as a result of western assistance efforts and the heightened sensitivity of local nuclear officials to safety problems. Some Russian efforts have been made to reduce the positive power coefficient in RBMKs and to improve scram performance. At least one site is in the process of improving operator performance by use of a site specific training simulator. These improvements are in the right direction; however, these RBMKs will still not have the defense in depth which is provided by an integral containment structure. The RBMKs and early VVER model 440/230s depart so far from what is now accepted in the West (e.g., lack of containment) that the general view is they cannot practically be upgraded to meet contemporary standards for safety and should not be operated any longer than necessary. The risk of a serious accident from these plants is considered too high.

Despite these continuing concerns, there is no evidence that the 1992 G-7 recommendation to close down the least safe plants (including the RBMKs) is being followed. Instead, it appears increasingly likely that Russia will continue to rely on their RBMKs for the indefinite future. The U.S. and others have started a successful program to equip these plants, as well as some remaining VVER 440/230s, with requisite upgrades to prevent severe accidents in the short term. The U.S. continues to believe high priority should be given to safety improvements to RBMKs and VVER 440/230s in Russia.

In addition to their design deficiencies, the safety of many of these nuclear power plants continues to degrade as a result of economic dislocations and political and structural problems following the breakup of the Soviet Union, which has resulted in a loss of centralized technical and regulatory authority and less financial support for nuclear power.

While there is broad agreement in the Western nuclear community on what needs to be done to address Russian nuclear safety issues, until very recently there has been no similar consensus on the Russian side on the most urgently-needed near-term assistance measures. Although some Russian officials agreed that substantial improvements were needed, the powerful Ministry of Atomic Energy (MINATOM) has rejected Western judgments that the RBMKs and VVER 440/230s pose sufficient inherent risks to be promptly shut down. Technological pride comes into play as does the fact that these facilities provide large quantities of energy the Russians argue is vitally needed. Additionally, although Russian officials do not often admit the point, foreign exchange earnings from electricity exports are also a factor in their reluctance to close down the oldest reactors.

Another issue is Russia's traditional nuclear power priorities, which have focussed on continued plant operation, and a consequent lack of attention to what the West considers to be needed near term safety actions. One result of such priorities was clearly seen in December 1992 when Russian Prime Minister Viktor Chernomyrdin issued a decree announcing an ambitious Russian program for doubling nuclear energy's output of electricity in the next 20 years. A blueprint for increased reliance on nuclear power in Russia's energy picture, it paid scant attention to improving safety at operating plants and did nothing to develop longer term alternatives to the RBMKs. We believe this failure reflects in large part the absence of an aggressive, independent nuclear regulator.

THE SOLUTION

The G-7 Approach

There is broad agreement in the Western nuclear safety community (formalized in a program adopted at the 1992 G-7 Economic Summit) on a three-stage approach for addressing Soviet power reactors.

First, **near-term** assistance measures must be taken to reduce the immediate risks of catastrophic accidents at the worst plants. The G-7 Program of Action called for immediate measures in:

- operational safety improvements (including training of personnel, upgrading fire protection systems and improved instrumentation and control);
- near term technical improvements to plants; and,
- enhancing local regulatory regimes.

A number of these improvements have already been started, and the U.S. is beginning to see results on the ground.

Second, a **mid-term** set of actions were deemed necessary to improve the better reactors (VVER 1000s and 440/213s) to enable them to operate safely for an extended period. Third, the G-7 Program of action sought to create the basis for **longer term** safety improvements by examining the scope for replacing less safe plants by the development of alternative energy sources and the more efficient use of energy; and, the potential for upgrading plants of more recent design.

The costs of upgrading the better plants will be large, potentially involving replacement of some 10-15 gigawatts of electric power, and consideration of funding is needed (a combination of some grant aid, with large amounts of commercial financing is probably needed). The G-7 advised that, to establish a suitable basis on which decisions on upgrading the better plants, long term plant safety, energy policy, alternative energy sources, and financing can be made, the following measures were needed:

- the necessary safety studies should be done, and
- the World Bank and the International Energy Agency should prepare energy studies that include assessments of alternative sources of energy and their cost implications.

Success in moving to the second and third phases of the G-7 program presumes major structural changes in the Russian and Ukrainian economies and in their energy policies, including market pricing of energy, demand reduction, conservation measures and alternative power sources. The efforts should be part of a market-oriented reform of energy policies encouraging commercial financing for the development of the energy sector.

THE U.S. BILATERAL ASSISTANCE PROGRAM

The political turmoil in the former Soviet Union in the past three years resulted in the political restructuring of the region that was more amenable to receiving aid from the West. But it also had the effect of disrupting maintenance and operational safety activities in the nuclear power and safety infrastructures in those countries. And it has resulted in a decentralized nuclear energy structure that has, to some extent, demoralized the design talents and technical knowledge in the former Soviet republics that is needed to support plant activities in countries outside of Russia.

The changes also led the NRC to alter its bilateral structure for providing assistance to Russia and Ukraine. Whereas we previously co-chaired a U.S.-Soviet Joint Coordinating Committee on Civilian Nuclear Reactor Safety, we now have joint committees with both Russia and Ukraine, and hold joint annual meetings. There are now both NRC and DOE Co-Chairmen, 2 Russian and 2 Ukrainian Co-Chairmen, making our coordination job more difficult.

With the high priority being given to the various Summit proposals, including the Lisbon Initiative, our Joint Committee work was recently limited to working groups in:

Reactor Vessel Radiation Embrittlement and Annealing
Severe Accidents
Environmental Transport of Radioactive Material and Health Effects (DOE), and
Nuclear Plant Aging and Life Extension

Under the Lisbon nuclear safety initiative the U.S. plans to assist in establishing regional training centers in Ukraine (Khmel'nitskiy) and Russia (Balakovo). DOE also has a program of operational safety enhancements and risk reduction.

NRC continues its program in support of better nuclear safety regulation. We are, for example, training former Soviet Union regulatory representatives to set up a governing structure adopting appropriate U.S. standards, requirements and procedures. One of our goals is that they will participate in simulator training and technical assistance areas developed under the DOE operational safety enhancement initiative. We also select key regulatory officials to participate in training in the United States and to observe planned team inspections. We are also providing assistance to establish a regulatory training program in Russia and Ukraine and to help them establish incident response programs including a response center in each country with communications capability to each power plant. We are providing fire hazard analysis and methodology training, implementing analytical methodologies in performing safety analysis of operating plants, providing assistance in the creation of a legal framework for regulation, and may help to develop their waste, spent fuel and nuclear materials regulatory programs.

In addition to these safety programs we are working with Russia, Ukraine, Kazakhstan, and Belarus on programs on physical protection regulation and on safeguards and non-proliferation regulation (which is supported under the Defense Department's Safe and Secure Dismantlement (SSD) or Comprehensive Threat Reduction (CTR) Program). There is a fully funded program on enforcement regulations. We are also advising drafters of regulatory legislation for Ukraine and Russia. Finally, FY 1993 funding is expected to assist in the procurement of VVER-1000 analytical simulators for Ukrainian and Russian regulators.

In a July 1992 meeting with the Chairmen of Russia's GAN and Ukraine's SCNRS (their respective regulatory organizations) we jointly established projects for assistance under the Lisbon Initiative. A December 1992 meeting with GAN developed implementing plans, and a March 1993 meeting with the Chairman of SCNRS furthered the Ukrainian program. About 60 groups comprising about 140 Russians and Ukrainians will come to the U.S. during 1993 to participate in these projects.

A JCCCNRS Working Group has also been working on cooperative efforts on health and environmental effects from the Chernobyl accident. A program to preserve the Chelyabinsk data and to undertake related scientific research is being pursued as part of a recently negotiated and initialed Radiation Effects Research agreement between Russia and the United States. Considerable understanding and cooperation have developed and resulted in excellent working relationships, benefitting Ukraine, Russia and the U.S.

Some of our cooperative efforts have developed into research projects sponsored and funded by NRC, but performed in Russia. Notable has been severe accident research in the areas of corium interaction with the vessel, hydrogen combustion and detonation, and validation of the CORCON-MOD 3 computer code. Recently, the Russians delivered a presentation to the Nuclear Energy Agency on their in-vessel melt retention experiments, asking NEA that this program become an international cooperative project between Russia and NEA member countries. NRC supports this.

In general, we have received very positive feedback from our Russian and Ukrainian counterparts with regard to the progress and direction of our efforts. The U.S. believes future assistance should concentrate on fire safety, operations, and design specific and plants specific modifications. We'd like to support Russia in performing an Individual Plant Examination (i.e., a systematic plant safety evaluation based on probabilistic techniques) on a VVER-1000 to establish a "model plant" safety analysis, implement short-term fixes on RBMK's and VVER-440/230's, and involve Russian and Ukrainian engineers in design, analysis and plant modifications.

AN INTERNATIONAL NUCLEAR SAFETY CONVENTION

We and others in the west are actively seeking to establish an international nuclear safety regime, but a number of countries, especially in the former Soviet Union, have had a tradition of giving priority to energy production rather than to safety. They also have weak or embryonic regulatory authorities and are operating plants with fundamental design weaknesses. In part for these reasons, and partly because a number of countries in Western Europe feared another serious nuclear accident in the East, an initiative was launched at the IAEA in 1991 to establish an International Nuclear Safety Convention, to codify certain basic principles of safety to which all countries could be held accountable. The United States supported this idea as one way to bring international pressure to bear on those countries where an inadequate safety culture increases the possibility of another major accident. We continue to believe the completion of the Safety Convention is in the interest of all countries and will be an important step towards assuring the safety of nuclear power plants throughout the world.

The elements for such a Convention have been under development for almost two years by a special expert working group sponsored by the IAEA. These technical discussions have proceeded slowly and in an uneven fashion, with significant disagreements on such fundamental issues as the scope of the Convention. However, recently substantial progress has been made towards a convergence of views, giving renewed hope for the successful completion of these negotiations. In fact, many are hopeful that an International Nuclear

Safety Convention could be negotiated at a diplomatic conference early next year, and opened for signature at the IAEA's September 1994 general conference.

The principal question at this point seems to be whether the countries in Eastern and Central Europe, and developing nations, will adhere to the Convention. Many of these countries have not been present in the negotiations or have participated only sporadically. It is unclear whether there are major political, technical or resource issues that could cause some nuclear countries not to join. As a result, we continue to place a high priority on identifying such problems and developing options for resolving them. To achieve its full benefits, the Convention must be broadly multilateral; in other words, it must have the widespread participation by countries using nuclear energy for the production of electricity. We remain committed to work actively to achieve this important addition to the international nuclear safety regime at the earliest possible date.

In Eastern Europe the NRC has been implementing A.I.D. sponsored assistance efforts amounting to about $1.6 million for FY 1991-92, with the Czech Republic, Slovakia, Hungary, Bulgaria and Lithuania, emphasizing computer code users group memberships, assignments to NRC, participation in western technical workshops, and training assignments at the NRC Technical Training Center. DOE assistance to Eastern Europe amounts to about $2.4 million per year in bilateral and IAEA-related safety activities.

MULTILATERAL EFFORTS

Clearly much is being done at the very highest levels in our own government and elsewhere to help countries using Soviet-designed nuclear power plants improve their safety. In addition to what has already been outlined, the IAEA continues its Central and Eastern European Safety Program. Their VVER 440/230 safety analysis work is virtually completed, while the VVER 440/213 and 1000/320 work is just beginning.

The IAEA has recently put out a report on RBMK safety, so analytical work is also beginning on this reactor. "Extra-budgetary contributions" of $1.5 million have been requested for both VVER and RBMK work. Meanwhile, the International Nuclear Safety Advisory Group has completed a revised assessment of the Chernobyl accident, emphasizing design issues, as opposed to the earlier report that placed most of the blame on the operators. Coordination is also improving with the IAEA and the European Commission Secretariat exchanging staff members to improve information flow.

The OECD Nuclear Energy Agency convened a meeting of senior regulators who asked for the NEA's views on assistance being provided from a regulatory perspective. Their meeting resulted in the establishment of a link between Western and VVER reactor regulators. Regulatory bodies from countries that own Soviet-designed VVER reactors are organizing once again to exchange views on regulation of these reactors, much as they did before the massive break-up of the Soviet sphere of influence.

All major nuclear countries are giving considerable attention to nuclear safety problems in the FSU and Eastern Europe. Several new actors are now involved in providing or coordinating assistance (G-7, G-24, EBRD, and more individual countries), and there have been multiple entry points for this involvement (for example, Sweden in Ignalina, France at Kozloduy, and Finland with Sosnovy Bor), with little substantive, advance coordination. Much useful work has been done on the ground, but there is still much left to be done, both in the regulatory context and in terms of energy economics. The NRC will continue to play its part in this coordinated effort.

Mr. SHARP. Thank you very much.

I think from both the things I have heard and the things you have said today that I know all of you folks and the people that work with you have, indeed, been diligently working on this issue. I think generally the American people might be very pleased to know that, while this may not always be on the front page in our news media, it has certainly been a very high priority begun by the last administration. Of course, I think we all recognize that the Europeans are involved, both where the plants are located and where neighbors have a profound interest and concern in this.

I guess the question I would start with is to try to get a little bit of grip on what the bottom line is at this point in terms of the highest risk reactors, and this partly goes to the recent Ukrainian Parliament decision. I think all of you indicated there is a consensus they ought to be closed down as soon as possible and, of course, the question becomes, what is as soon as possible. Now we see that there will be a delay in the Chernobyl plants being closed, which I understand everybody agrees are the highest risk, the RBMKs. Those are the highest risk plants.

I wasn't sure whether I detected any difference between Ms. Verville and you, Dr. Selin, in terms of this question of what short-term steps to take at these plants, the implication being, not much to try just to get them shut down, or whether there are things that we ought to do?

Mr. SELIN. We have to be realistic. No matter how much we may agree in this room that we would like to see these plants closed down, they are going to continue to operate and they are going to continue to operate for a while.

The short-term strategy has two possible benefits. The first benefit is to reduce the risk of these plants as they operate, and there has been significant reduction in risk. I don't think the Americans can take the major part of the credit. The credit must go to the Russians, primarily to the Russians but also to some of the other countries for seriously addressing the safety risks and at least attacking the short-term question, the safety culture, operational questions, the technical improvements. So we are pleased to see some reductions in the short-term risk of these reactors.

At the same time, as all of you have pointed out, we have a role to play in trying to convince the Russians that in the mid-term and the longer-term it is not in their interest to continue to operate these plants. This is very difficult to do and it is expensive to do. We can't just exhort them to close down the plants, we have to work with them at alternative sources of electricity since the economics and the safety are inextricably intertwined. But we don't have standing or leverage, at least we didn't a year or two ago.

Having done effective work, even the high office told the European parliamentarians that the American program is more effective than theirs. I think he did it not as a compliment to us but to catch their attention. Nevertheless, the effectiveness of these programs and the building up of the confidence gives us at least the basic standing to engage in these much more difficult long-term arguments.

The arguments are very difficult and it is going to be very tough to make progress in both convincing the Russians and in finding sources of finance to deliver on the arguments for replacements. These replacements might be finishing some of the better Russian or Soviet-designed plants that are under construction in Russia and the Ukraine. It might involve some of the advanced coal-fired technologies to use coal. It might involve some gas work. But the Russians are clearly not going to close the reactors until or unless they have some alternatives.

We now have the basis for discussing with them as proven friends instead of just critics from the West what this basis might be. It is going to be hard to convince them and it is going to be hard to finance the results even when it is done, but that is the program which the administration has now launched.

Secretary O'LEARY. I would just like to add some focus on the question of projects alternative to nuclear generation. One of the things we are focusing on in the fossil energy area is work on gas turbines, which is one of the demonstration projects which comes out of agreements signed in September. There is also one clean coal technology project, and another project with respect to an electric power plant.

I agree with Chairman Selin and would like to put in the discussion a little personal experience which, again, came out of the trip I took to Russia last month, and that was clear discussion, both directly to me and in side conversations which was not specifically to be translated to me but was translated to me, I got the certain impression that our short-term goals do not at all match the Russian long-term goals. If one examines the economics of a country, it becomes clear why not, and that is simply, with hard currency in such short supply, the only alternative that they now see and can focus on is sale of the natural gas, which we now look at as the alternative to supplying the energy desired to back us off of those less safe nuclear power plants, which they see as a commodity to be moved to the West or the East. It is that dilemma, the need for hard currency and the markets available coming out of the alternatives that make this such a difficult task.

The other point I would like to bring to the committee is that in meetings that I held, both bilateral meetings and multi-lateral meetings in Paris at the International Energy Agency as well as in Vienna at the International Atomic Energy Agency, it is clear that our Western partners see us as the moral leader on this issue.

I think we have a responsibility to continue to speak on this issue, but it is these international fora where there is the combined voice of both the United States and all of Western Europe sitting in that ring where the dangerous or the least safe plants exist, which will get this point finally brought home. But, it cannot be acted upon until there is the economic wherewithal to do so.

So it will be the funding, it will be the private sector interest, and finally the moral suasion of the West that will get this piece done. I think any short-term suspicion that we are going to make progress will need a lot more muscle put to it, and that is just, once again, very blunt O'Leary fashion.

Ms. VERVILLE. I just would like to confirm, in case there has been any misunderstanding, that I certainly agree with both Chairman Selin and Secretary O'Leary. I would simply add that the efforts I was describing in the Group of Seven Nuclear Safety Working Group coming out of the 1993 Munich Summit are also very much focused on looking at this medium and longer-term and the question of possible alternate energy sources and what resources are available in the international community to put together behind the moral suasion in order to try and look at this issue country-by- country to see what might be realistic. There is recognition that it is a very big task, but there are mechanisms underway to try and address it and to try and work both bilaterally and in these various multi-lateral fora.

In other words, we are trying to get our act together.

Mr. SHARP. Sure. I certainly appreciate that, and I certainly appreciate the effort that people make on a daily basis without both sometimes the achievement that helps motivate or there are outside forces that help to motivate.

I guess I am trying to get an impossible grasp of the intensity of concern here, not only your part, the intensity of concern because I certainly can understand all the institutional and economic and political turmoil. If these things aren't a profound risk to you immediately, they are not going to get in a political and economic climate as it faced in these countries, I think it would be extremely unrealistic to think they were going to get the attention that one would like to give them unless they are seen as an extreme risk.

Partly, I would ask whether you see them seen that way by their neighbors?

In other words, in an immediate threat sense, we have less at stake as to what happens in those reactors than do people that live within so many miles of them obviously do kind of things, and I guess I would ask whether you think the other Western European nations which we have heard from before on this subject really see it that way or whether it is sort of handwringing and then everybody says, "Well, do the best you can," and that is maybe the best that is going to happen?

Secretary O'LEARY. Clearly they see it that way.

Mr. SHARP. See it that way as the serious threat?

Secretary O'LEARY. See it as a serious problem and it is not simply a handwringing after dinner. It is in every bilateral conversation that I have had in the 9 months that I have been in this job. Anybody sitting on the ring of the former Soviet Union, or West Germany and now the unified German countries, the Swedes, the English, and the French, all not only see this as a critical issue but, in point of fact, they have begun in a very real way to support these efforts in terms of the work we have been doing and the funding provided.

The difficulty that we have—and this is interesting because here I am the newest person on the block now speaking about this so passionately, so I will shut up in a minute and let others do it much better—is that we are all seeing the problem in the same way. Now the question is, how in a land where people have failed to come to grips with this issue do you bring home a truth that often is very difficult to accept. That is why we have to keep pounding at it and keep working at it.

Ms. VERVILLE. I would just add that certainly at the head of state level in the Group of Seven, nuclear safety has been an important agenda item the last 2 years running, and with a coherent action plan to look both at the short-term safety problems and the longer-term issues that you are speaking of now. The G-7 Heads of State have charged their countries to work with these multilateral institutions that have more money than the individual countries.

Nuclear Safety has been a focus of high level attention in all these countries. The Western European countries, for example, were the prime motivators behind setting up the nuclear safety account at the EBRD which has had, as I said, the first success——

Mr. SHARP. And what is the account going to pay for, is it going to replace the reactor or is it simply——

Ms. VERVILLE. The G-7 has charged the account to look at risk reduction measures in the short-term, but in so doing the EBRD staff has been working with the World Bank, the Group of 24, and the other institutions that have been established in order to try to get a commitment for early closure of plants that are of the riskiest nature.

In fact, in the first project, which is the only one to date that has actually been fully developed, for Bulgaria this has been achieved. So it is a start.

Mr. SHARP. In other words there is at least partial funding from outside of Bulgaria to close the plant?

Ms. VERVILLE. The nuclear safety account which has contributions now of about $150 million from Western countries, is funding, I believe it is $25 million, or for upgrades at the riskiest plants in Bulgaria. This is in combination with World Bank and other EBRD lending to bring on line alternatives so that the Bulgarians are in a position to make a political commitment to early closure of the riskiest units. It is a start.

Mr. SHARP. I see. In other words, the goal is, get them committed to the closure of that and the money will help finance alternative sources of energy plus temporary upgrades?

Ms. VERVILLE. The money will be used for short-term safety upgrades. We have to accept that the plants will continue to operate for the short-term because they are needed. The goal is to put short-term upgrades together with longer-term planning for alternatives, in order to get a political commitment for early closure of the riskiest plants.

I think that other Western governments are very serious about it, certainly in the fora which I attend.

Mr. SHARP. I can appreciate that those are extremely important and valuable steps to take. I am not belittling or doubting that at all, but I sort of put the test as to how people view a risk as to whether they will put their money where their mouth is.

Mr. SELIN. The Europeans through 1992 have put up more money than we have. They have centered it more in Central and Eastern Europe than in the Russian Federation. The problems are more tenable and the money hasn't been spent that productively, but they have been learning and figuring out how to spend this money. Bulgaria is a relatively easy job because they are also bringing online two of the 1,000 megawatt reactors, so quid pro quo is feasible there.

Talk a little bit about attitude, I have watched this for a long time and I think the Russians did nothing between 1986 and 2 years ago, or at least there was no practical impact. But, I think, in the last 2 years there has been a significant improvement in short-term risk reduction. The West has put up a fair amount of money, but we have not been the leaders until about a year-and-a-half ago in either funding or taking the leadership.

The key thing is that the risk reduction measures are foreseen as being funded through grants whereas the longer-term measures clearly have to be upon a commercial basis. In order to be able to pay for the longer-term investments, whether it is gas-fired or coal-fired or even the better nuclear plants, there have to be major economic reforms as well as the technical safety reforms. The key thing is that electricity has to be priced at a level that will produce enough income to cover the investments and to amortize the loans that we are talking about.

I know it is a little weird for the safety regulator to be saying that if anything economics are more important than the technical questions in Russia, but it really is true. We are not going to put in $50 or $75 million.

Mr. SHARP. The safety requires the capital?

Mr. SELIN. Absolutely.

Mr. SHARP. Let me apologize to my colleagues. I got carried away and forgot they were here.

The gentleman from Massachusetts.

Mr. MARKEY. The whole planet could get carried away with this one.

Condition 8 of the Senate's resolution of ratification of the Start 1 Treaty, the so-called "Biden Condition" directs the President to seek monitoring arrangements for warhead and fissile material inventories in connection with the implementation of Start 2.

Given the continuing turmoil in Russia and the other states of the former Soviet Union and our lack of knowledge about the disposition and size of the Russian nuclear weapons and material inventory, what is DOE doing to accelerate decisions on the release of nuclear stockpile data that could be exchanged with Russia on a bilateral basis?

Secretary O'LEARY. We have had a series of discussions with our counterparts in Russia on the whole issue of exchange of data, again growing out of the Gore/Chernomyrdin Commission. Those issues were discussed for the first time formally in September and, subsequently this month, and there will be continuing discussions as we move into November and December. To promise you that we will have something to report prior to our December meeting, I think, would be foolish.

Mr. MARKEY. Did you set a deadline for coming to a resolution of the issue?

Secretary O'LEARY. We have not yet set a deadline. I think that will happen in December or prior to December.

Mr. MARKEY. What is the timeframe that you are talking about generally now in terms of the exchange of information, is there a parameter that you have built around the discussion?

Secretary O'LEARY. We are talking about the earliest possible exchange of information and, with respect to specific dates, I am not prepared to provide them until after the next series of meetings.

It is clear to me that on both sides there is a desire to begin an early exchange of data and, from the Department's perspective, we will have some announcements to be making with respect to this issue early next month. Perhaps on the fourth, if not on the fourth on the tenth. With respect to the Russians, again, I would have to provide that information after the next round of meetings.

Mr. MARKEY. As you may know, the House version of the defense authorization bill includes an amendment that I offered that limits the expenditure of the United States Nunn-Lugar funds for a surplus plutonium facility in Russia until the President certifies to Congress that Russia is committed to halting the chemical separation of weapon-grade plutonium and is taking all practical steps to end such separation at the earliest possible date.

A few weeks after I offered this amendment, the President announced at the United Nations that the administration would be seeking to negotiate a worldwide cutoff in the production of fissile materials for weapons purposes.

My amendment also expresses the sense of Congress that a key objective of the United States with respect to non-proliferation should be to obtain a clear and unequivocal commitment from the government of Russia that it will cease further separation of plutonium across the board, both military and civil, not later than June 1st, 1994, the President must submit a report on the status of U.S. government efforts to achieve these objectives, including the status of joint efforts between your department and MINATOM to shut down the remaining Russian dual-purpose production reactors or convert them to a non-weapons plutonium cycle.

In line with the President's announced policy and the thrust of my amendment, what can you tell me today about the status of your department's efforts to shut down or convert the remaining Russian production reactors at Tomsk and Kvasmoyarsk, what have you proposed to the Russians, how are they responding and when might we see results?

Secretary O'LEARY. I can tell you that our lead laboratory on this effort, which is Pacific Northwest Laboratories, has been working on this issue with the Russians for practically 1 year. We are well aware of the deadlines that have been established in the amendment. They have told me to tell you, as I have told you earlier without prompting, that these discussions are ongoing and I suspect that we will have something to offer to you in the very near future, and I have set the timetable for December.

We don't take this instruction lightly. It was one that we had taken on for ourselves a year ago and I will do all possible to ensure that the deadline is met. You can be assured that the President will do all to assure that I make that deadline.

Mr. MARKEY. As you know, the amendment fences the money that would be used for plutonium storage, several hundred million dollars, which would go to the Russians otherwise, unless they commit to stop providing the production of new plutonium for their reactors.

Secretary O'LEARY. Yes, I understand that.

Mr. MARKEY. I would hate to have this thing next year be a major debate out on the floor, and I would just urge you to negotiate a resolution of it that satisfies the desire of the American people to help but, on the other hand, not have storage of the old fissile material while they are generating new unneeded material at the same time.

Thank you, Mr. Chairman.

Mr. SHARP. The gentleman from Texas, Mr. Barton.

Mr. BARTON. Thank you, Mr. Chairman.

I want to thank the distinguished panel for appearing.

My first question would be probably directed to the Chairman of the NRC, what is the risk of an accident in the worst Russian reactor, one in a thousand, one in ten, one in a billion, what is it?

Mr. SELIN. It is very hard to put a number on this, but it is probably on the order of one in a thousand. In other words, one chance of an accident at each of the Russian reactors in a thousand years, it might be worse than that.

Mr. BARTON. How would that compare to an American reactor?

Mr. SELIN. It is at least a factor of ten worse than the American reactors.

Mr. BARTON. So their worst reactor——

Mr. SELIN. They have 15 of these worst, between the Russians, the Lithuanians and the Ukrainians.

Mr. BARTON. They are a thousand times more likely to have an accident?

Mr. SELIN. Each one is about the order of ten times more.

Mr. BARTON. Ten times?

Mr. SELIN. Yes, sir.

Mr. BARTON. In terms of the money that is being spent on risk reduction and policy changes, are the former Soviet Union republics putting up any money or are we putting up all the money?

Mr. SELIN. No, sir. In fact, it is impossible for us to know what they are putting up in the way of cash, but we can look at what they are funding versus what we are funding, and the efforts of the West to date on the RBMK, the Chernobyl-type reactor has really been quite modest. There has been a lot of analyses, there have been some investments in fire protection, but the actual expenditures have been almost all Russian financed to date. I think we can take some of the credit for increasing their sensitivity and working together with the Russians to point out which areas would be most productive, but there have been major investments—on the orders of tens of millions of dollars if they had been in U.S. terms—per RBMK reactor in improving the material condition, the physical part of the plant.

I would cite three areas. One is that the control rods which were used at the time of Chernobyl were both very slow, and they had the very unfortunate characteristic of first speeding up and then slowing down the reaction. These apparently have been replaced. They seem to have been replaced in all of the eleven Russian reactors, and I think in the two Lithuanian reactors as well.

The second is they are using a more highly enriched fuel which greatly decreases the runaway characteristics in certain circumstances of the RBMK.

The third is, they have added a fair amount of computer, fairly rough, but computer capability to the control room.

So they have made some significant improvements, but they don't address the fundamental problems in the reactor. They are dealing with what they can deal with, not with the things that are essentially not solvable and for which reason we, therefore, wish to see the reactors closed.

Mr. BARTON. I know it is a longer-term issue, but is there any negotiations or discussions and is it any interest on their using American reactors in the future instead of what they are using today?

Mr. SELIN. Not reactors, but there is some in components. There is a very important program, not in Russia, but in the Czech republic in which the Czechs have selected equipment from Westinghouse, information instrumentation and control equipment to control not one of the higher risk but the most modern of the Russian reactors, the 1000-megawatt pressurized water reactor.

In order to provide this equipment, Westinghouse has had to do a thorough analysis in great detail of exactly how the reactor would respond to safety challenges in a large range of conditions, and what control measures should be taken.

They and we, the U.S. Government—not we the NRC—are quite comfortable that this would be a happy marriage, that the Russian-designed Czech-built system with the American instrumentation and control would turn out to be quite an effective reactor. We hope, I can't say that we have seriously discussed this with the Russians, but we would hope that this would be at least one of a number of choices the Russians would consider further down the road when they have generated enough cash to look at some serious improvements in the longer-term.

Mr. BARTON. Would that component system that you just mentioned, would that meet American safety standards?

Mr. SELIN. There is no intrinsic reason that the most modern Russian reactor could not be brought up to American safety standards. They would have to do a lot of things. One of them would be to have a modern instrumentation and control system. So in that narrow area, the 1000-megawatt reactor plus the instrumentation and control system would meet Western standards.

Mr. BARTON. If they came to your commission for a license to construct and operate, then you would give it?

Mr. SELIN. They would have to do a lot of other things, but there is—I will turn it around negatively, we would not look at their instrumentation and control, the control of the reactor between Russian design, Czech construction, and American instrumentation and control, we would not say that failed our standards. There are a lot of other things they would have to do that are not affected by this set of equipment that they would also have to meet. If they met those, the 1,000-megawatt reactor with this change and with a number of other things would intrinsically be licensable by American standards.

Mr. BARTON. It intrinsically would be licensable?

Mr. SELIN. There are a lot of things that have to happen in terms of operation as well as in terms of the physical condition of the reactor.

Mr. BARTON. I was going to ask about coal-fired and gas-fired plants, but Secretary O'Leary basically answered that question in response to another question.

If you were the chairman of the equivalent to the NRC in Russia or the former Soviet republics today and had the same licensing standards that we have in this country today to operate, how many of their reactors would remain in operation?

Mr. SELIN. None.

Mr. BARTON. You would shut them all down?

Mr. SELIN. Maybe two. There is a reactor in Kazakhstan that probably could meet our standards, and there is a like one in Russia I haven't visited that might meet our standards.

Mr. BARTON. But basically their entire nuclear power generation system doesn't pass muster as we think it should?

Mr. SELIN. That's correct, sir.

Mr. BARTON. I guess my final comment is to the Secretary. I want to thank you for the support that you have shown through difficult times in the last several weeks on another issue.

I would just, since you had some discussion with some of their senior officials, do they see any sense of urgency in addressing the safety problem that the hearing is about, or is it something that they have so many other alligators that it is kind of a second tier priority with them?

Secretary O'LEARY. I think the chairman has expressed it well with respect to the change over the past 2 years with respect to addressing short-term safety issues.

I would like to discuss another personal experience and tell you about what I understand about some of these operators in these plants. Before I came to work for this government and I worked for a diversified utility, one of the things we had done under INPO—and now I owe the quarter—was to enter into some partnerships with utilities both in what is now the Russian Federation and also with Lithuania. I can recall back as early as 1982—I am sorry, '91 and '92. Russian plant operators coming to the United States to Minnesota to work in exchange and learn using our simulators and, in some instances, these operators were onsite and at the plant for 6 to 7 weeks.

Clearly that they felt honored to participate and, more importantly, what I saw there was a display of pride in the sense that they understood what the requirements were in the United States and wanted also to learn and take those experiences back to their plants. I think that is a key to how we will come to the turnaround.

With respect to senior officials, I believe most are focused on the much longer term involved in economic survival for the next 5 years, and perhaps most rationally economic survival for the next 1 year.

That does not mean that these issues do not play. I mean, I have had very interesting and lengthy discussions with my counterparts, both in Energy and MINATOM on these issues. I think they see them as necessary. But again, as I indicated before you arrived, our timetable and theirs are not in sync. Our time horizon is much shorter than theirs. It is there that we need to apply the work.

Mr. BARTON. It's a concern. I'm glad to know that they are investing some funds, as the chairman said, and that they have shown some sensitivity. But at the same time, we have a little clip here that shows the Russians, last week I think, dump low-level nuclear waste off the coast of Japan—not once, but twice.

You say once. The report indicated twice. So maybe it wasn't. I will take your word for it.

Secretary O'LEARY. Never believe the press.

Mr. BARTON. I know. We live here in a country where people are concerned, and rightly so, about the environment and the cataclysmic calamities that can occur if nuclear power is not properly maintained. Certainly the chairman has been an advocate of the strictest security, which he should be.

But I don't sense that certainly in the former Soviet Republics that they have that same sensitivity. It is because putting bread on the table tomorrow is a little bit more of a predicament than worrying about a one in 10 chance of an accident, or whatever it is, in their reactors.

With that, I would yield back to the chairman. That is just a comment. You can comment on the comment.

Mr. SELIN. I would just say that although I agree with your observation, the problem is significantly more serious than that. It is not just economics. The Russians have produced more atomic contamination of their land than everybody else put together.

It is a long tradition of absolutely cavalier treatment of the environment. Nuclear safety is only one aspect of it. Dumping is another one. It takes a long time to change these attitudes. The Russians see these short-term improvements as final steps in themselves. To us, they are just the ante to get into the card game.

So whereas we may have some agreement on what needs to be done in the next couple of years, as the Secretary said, we see this from a very different framework.

Mr. BARTON. I yield back.

Mr. SHARP. Thank you.

Let me ask you this question. Obviously the Soviet Union produced extraordinary engineering and scientific talent that developed a host of things—from science, the space programs, to the whole nuclear program that they have.

How much risk is there to the people working at these various sites? Is that substantial in terms of their personal health? Is it substantial enough that aside from economic considerations, personnel are harder to attract to do what needs to be done? Is that a growing risk now that the information is less controlled in the former Soviet Union? Do these people see there is no problem?

I hear there are even nuclear reservations. They may not be the power plants—I don't know—where Western experts, when touring, refuse to either spend much time or even go on site; is that correct?

Mr. SELIN. Look at the bottom line. Things have changed enormously, and mostly for the worse, for people who work in these nuclear facilities.

Mr. BARTON. Right.

Mr. SHARP. During the Soviet Union days they were the cream of the crop, whether it was the Chelyabinsk, or one of the power plants, or what have you. They were quite unsafe then, but people really would do almost anything to get these jobs. Perhaps the sensitivity about the security wasn't so high. But for many reasons these were highly desirable places.

It is very uneven now. I don't know the weapons facilities, but I do know the power plants. Every one of the power plants retaining people has become a major problem. These are remote towns. They build a stand-alone city with a power plant around them. They used to get the top draw in supplies. They are having trouble doing this now.

The pay is uneven. The companies that buy the electricity are very uneven and the time in which they pay for the electricity. So, having cash, whether it is to pay people or to get supplies, has become difficult.

But the people are still running the plants. There are sources of concern but if you go on the site, they are fully staffed. The people seem to be fairly competent. It is a serious problem. I think it is will more serious rather than less in the next year or two. But the retention of personnel, at least in the power plants, doesn't seem to be in a crisis mode. The weapons facilities, I can't talk to. I don't know that myself.

Mr. SHARP. So you don't see it as a high probability that they would face the added risk that you can't keep on site talented people because it doesn't pay them, or doesn't have the privilege, or the safety to themselves is too great.

Mr. SELIN. I would put it differently. I would say that these economic reforms which allow cash to be retained at the plant, not soaked up in Moscow, which allow long-term budgeting, which allow proper payment of people, that allow the building of a sinking fund, to amortize capital investments—these are absolutely prerequisites for safe operation.

If that doesn't happen, you will see the effects reflected in many ways, including the lack of retention of personnel.

But I think it is solvable with money. I don't think the safety issues have been such to discourage people from working at the plants. I mean, in 1957 you had the worst accident in the history of the world, the worst nuclear accident at Mayak. People continue to work there.

Mr. SHARP. Do some of us overestimate those risks then? I realize we have a standard of risk.

Mr. SELIN. I don't see any experts on Russian psychology at this table.

Mr. SHARP. I meant in our own society we see those as unacceptable risks. I don't know. Maybe we have become too spooked about them. I am just raising the other end of the spectrum now.

Secretary O'LEARY. Mr. Chairman, I think that two other issues should be raised to amplify on the excellent response of the chairman.

One is the concern with respect to the health effects of exposure. That is one that is important for us to understand in the United States, both with respect to the Russian population and, quite frankly, we would like to have a better handle on it here in the United States as well.

One of the agreements that has come out of this last round of the Gore/Chernomyrdin Commission meetings is an agreement that we will undertake a study to provide us with some real data on this issue.

I don't hold out the promise that this will be done quickly because, as you well know, there are very involved detailing of finding people, tracing their records and their history of exposure. The important thing to be said, though, is that there is agreement that this should be done.

The second issue, which I believe needs to be addressed is that of the plight of the scientists who have been involved in this work. This is especially true both with respect to the civilian plants, and certainly with respect to the weapons production sites. This is a major concern to us in the Department of Energy.

We are very close to an agreement, or had been before last month. But there should be established in Russia a center, which would provide employment for some of the scientists, who are coming out of the nuclear production plants and halt this flight to other countries other than the United States, where they cause a potential for nuclear proliferation.

In my last meeting—well, quite frankly, my only private meeting with the Prime Minister of Russia last month, we urged that either a decree or some action on the part of the current government, that the necessary work be done to quickly establish a center. It will be one way to hold the workers there.

The other opportunity to hold these scientists at work relates to what we are calling our defense conversion effort. This is yet another subgroup under the Gore/Chernomyrdin Commission. For the United States side, Deputy Secretary Perry of the Defense Department heads this effort. Victor Reis, who is our Assistant Secretary for Defense Programs, sits on that committee.

Our view is that this will be another way to provide meaningful work for some of the scientists, as well as for some of the people who have been working in manufacturing to move them out of this work. Once again, this is a long-term initiative, which addresses some of the issues that you have raised.

Ms. VERVILLE. Mr. Chairman, on the slightly more specialized aspect of the former weapons scientists, I am also responsible for International Science and Technology Centers in the former Soviet Union in Russia and Ukraine.

One of the agreements signed during Secretary Christopher's visit was with Canada, Sweden, and Ukraine to bring on-line a center in Kiev that would provide alternate civilian scientific employment for former weapon scientists in Ukraine. A likely area for potential projects and work, which would be with Western collaborators, would be the nuclear safety area.

In Moscow, the agreement has been signed for the International Science and Technology Center, in this case, with the European Community, Japan and Russia. It has been awaiting Parliamentary ratification.

I have just received word that we may have some information very soon from Russia about being able to start up the Center in advance of Parliamentary action in some fashion so that the extensive preparatory work which we have done, which includes potential nuclear safety projects, might be brought on line very quickly. We are very encouraged, and hope that this will happen very soon.

Thank you, Mr. Chairman.

Mr. SHARP. Well, thank you very much. I appreciate your time and attention to this for us today, but more importantly what you do year-round on this kind of a question because it obviously of considerable significance. It helps us get a better perspective on our efforts here at home that we make on safety, not to suggest we ought to diminish our efforts here. But it certainly does not look greener on the other side of the fence.

Thank you very much. We appreciate it.

Secretary O'LEARY. Thank you, Mr. Chairman.

Mr. SHARP. I need to put into the record the statement of Mr. Bilirakis and Mr. Moorhead. Their opening statements will be made a part of the record.

[The opening statements of Hon. Carlos J. Moorhead, Hon. Michael Bilirakis, and Hon. Gary A. Franks follow:]

STATEMENT OF HON. CARLOS J. MOORHEAD

Mr. Chairman, thank you for holding this hearing on the safety of Soviet-designed nuclear reactors.

As the 1986 accident at Chernobyl demonstrated, the threat of unsafe nuclear operations are not merely a local or national problem. The threat posed by reactors in the former Soviet Union and eastern Europe not only threaten the health and safety of their neighboring countries but pose a threat to the global environment.

Since 1986, we have found out a great deal about the unsafe design and operating conditions of Soviet-designed nuclear reactors. Now we must look toward encouraging those dependent on these types of reactors to begin to take these reactors off line or begin to improve the safety of these reactors' operations.

This year the United States is spending $30 million to improve the safety of these reactors. I believe it is important that we engage in these kinds of activities. I do, however, have some questions about how much money it will take to make these plants meet western safety standards and how effective our activities have been thus far. I also believe there are some innovative exchanges we could engage in with the states of the former Soviet Union so that we are not merely giving them aid but getting something of value in return.

I welcome each of the witnesses and look forward to hearing their testimony.

Thank you Mr. Chairman.

STATEMENT OF HON. MICHAEL BILIRAKIS

Mr. Chairman, I commend you for holding this hearing on the safety of Soviet-designed nuclear reactors. This is an issue in which I have long been interested, not only because of the risk of global environmental damage from sites in the former Soviet Union and eastern Europe, but because of the threat of these types of reactors being brought on line in Cuba.

The threat posed by Soviet-designed reactors is enormous. The foreign minister of Belarus (bell-uhroose), the newly independent republic just north of the Chernobyl site, recently announced that 800,000 of its children have been affected directly by Chernobyl. He also said 25% of the territory of Belarus is still moderately to heavily contaminated and 18% of agricultural land cannot be used. He equated this to a limited nuclear strike-carried out in peacetime.

This is a frightening statement for the people of Florida, who are only some 60 miles north of Cuba at Key West. The reason is this: in 1976, Cuba and the Soviet Union concluded an agreement to construct two nuclear power reactors near Juragua (your-ah-owa). Construction on the first Cuban reactor was begun in 1983 and on the second in 1985. I wrote a letter to Fidel Castro in 1986 expressing my deep concerns at the construction of these reactors.

Since the break-up of the Soviet Union, construction on these reactors has been suspended—and because of the threat that these reactors pose to the United States, and Florida in particular, I hope that construction never is resumed.

I do not say all this because I believe nuclear power is unsafe. In fact, I firmly believe in preserving the nuclear option and that nuclear powerplants, when properly designed, constructed and regulated, are not a threat to public health and safety. Florida is ample proof of this fact: four nuclear reactors have operated in Florida for many years without posing a threat to the environment or the public. Moreover, one of these reactor designs—at Port St. Lucie—is widely hailed as a potential standard for the industry.

Mr. Chairman, I look forward to hearing the testimony of our witnesses today. While I have serious questions about the safety of Soviet-designed nuclear reactors,

I believe that nuclear power must be a part of our energy mix—particularly considering our concerns with global climate change.

Thank you.

STATEMENT OF HON. GARY A. FRANKS

I would like to thank the chairman for holding today's hearing on efforts of the United States to improve the safety of Russian-designed nuclear power reactors. I look forward to the testimony to be presented today.

Recent reports indicate that the Ukrainian government intends to continue operation of the Chernobyl reactors. This decision is cause for great alarm, however, based on Ukraine's dependence on nuclear power for a significant amount of its electricity, one can see why this decision was made. However, this increases the potential for another Chernobyl accident and raises concerns of all officials trying to ensure a safe environment for nuclear power operations in this area. I applaud the efforts of this subcommittee to examine current efforts to improve the safety of nuclear power operations in the newly independent states of the former Soviet union and the former eastern block countries and look forward to the testimony to be presented today.

Mr. SHARP. The subcommittee stands adjourned.

[Whereupon, at 11:30 a.m., the hearing was adjourned.]

○

78-866 (76)

TP
APEX
9780160442599
○ ○ ○